Topic P1 — Energy

Topic P2 — Electricity

Topic P3 — Particle Model of Matter

Topic P4 — Atomic Structure

Topic P5 — Forces

Topic P6 — Waves

Topic P7 — Magnetism and Electromagnetism

This is a list of the equations you'll be given in your exams. You can look up equations on this list to help you answer some of the questions in this book.

Cells

Here we are. First page of the book. Good to see you looking so keen(-ish). Let's get started.

Warm-Up

Cells are the basic units that all living organisms are made up of.

Bacterial cells are different from plant and animal cells.

Tick the boxes to show whether the statements apply to plant, animal or bacterial cells. Some of the statements apply to more than one type of cell.

	Bacterial Cells	Plant Cells	Animal Cells
Has a nucleus.	☐	☐	☐
Contains a single loop of DNA.	☐	☐	☐
DNA is often found in plasmids.	☐	☐	☐

Bacterial cells do have some things in common with plant and animal cells. Circle three features of a typical plant cell that are also seen in bacterial cells.

chloroplasts vacuole cell wall

cytoplasm cell membrane

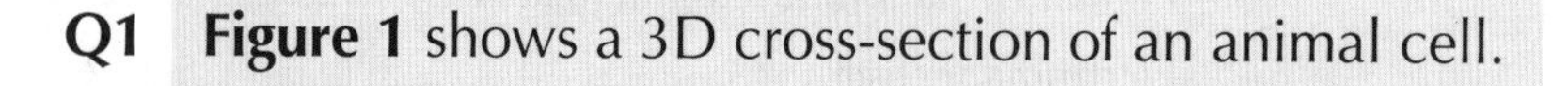

Q1 **Figure 1** shows a 3D cross-section of an animal cell.

Figure 1

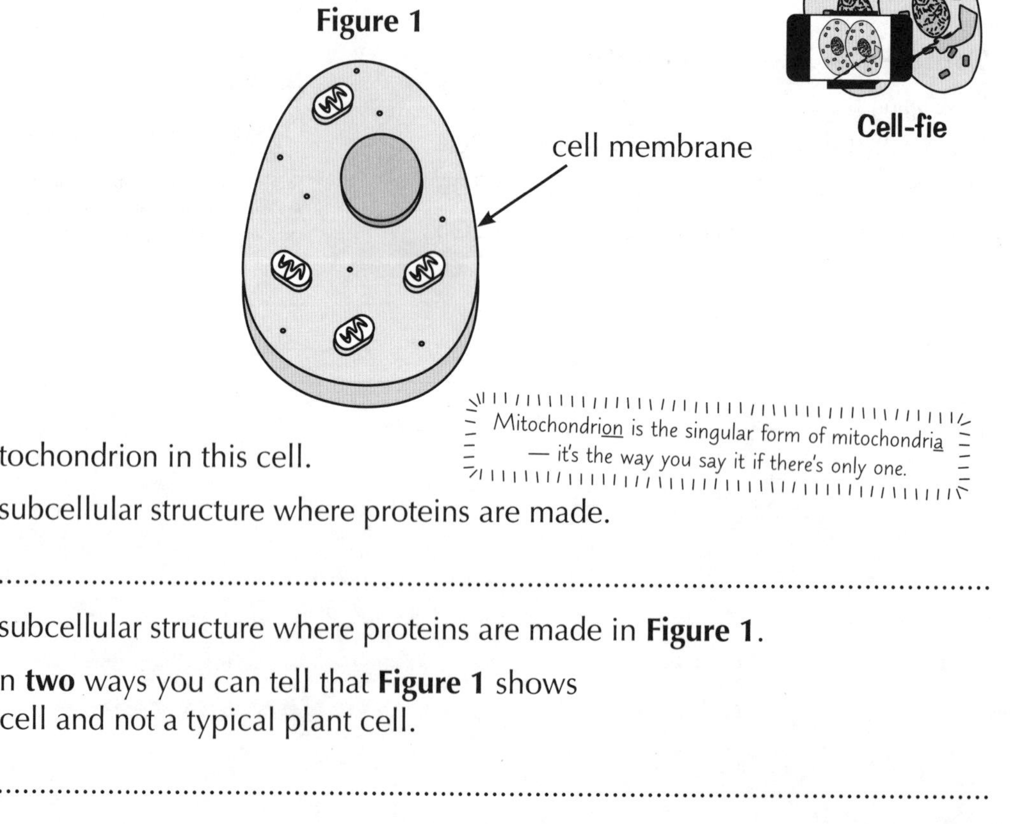

Mitochondrion is the singular form of mitochondria — it's the way you say it if there's only one.

a) Label a mitochondrion in this cell.

b) i) Name the subcellular structure where proteins are made.

....

ii) Label **one** subcellular structure where proteins are made in **Figure 1**.

c) Write down **two** ways you can tell that **Figure 1** shows an animal cell and not a typical plant cell.

1. ....

2. ....

Contents

Use the tick boxes to check off the topics you've completed.

Topic C1 — Atomic Structure and the Periodic Table

Topic C2 — Bonding, Structure and Properties of Matter

Topic C3 — Quantitative Chemistry

Topic C4 — Chemical Changes

Topic C5 — Energy Changes

Topic C6 — The Rate and Extent of Chemical Change

Topic C7 — Organic Chemistry

Topic C8 — Chemical Analysis

Topic C9 — Chemistry of the Atmosphere

Topic C10 — Using Resources

Q2 Read the descriptions of cells **A**, **B** and **C** below.

Cell **A** has a cell wall. It has no nucleus and no mitochondria.

Cell **B** has a cell wall and a nucleus. It also has chloroplasts.

Cell **C** does not have a cell wall. It has a nucleus and mitochondria.

a) Which cell (**A**, **B** or **C**) is a prokaryotic cell? ....

b) Which cell (**A**, **B** or **C**) is likely to be the smallest cell? ....

c) Cell **A** cannot carry out aerobic respiration. Suggest why.

Hint: think about the subcellular structures needed for aerobic respiration.

....

....

Q3 A scientist wants to examine cells from a bean plant under a microscope.

a) The scientist is looking at the cells from a leaf of the bean plant.

i) Name the subcellular structures in the leaf cells that allow the bean plant to photosynthesise.

....

ii) Photosynthesis produces sugars. Some of these sugars are stored in cell sap. Where in the bean cells is the cell sap found?

....

b) The scientist wants to investigate the cell walls of the bean plant cells. To do this, she needs to stain the cells to highlight the walls. **Table 1** shows a selection of stains and what each one is used to highlight.

Table 1

Stain	Used to highlight
A	DNA
B	Starch
C	Cellulose
D	Lignin

Which stain (**A-D**) should the scientist use to investigate the cell walls?

....

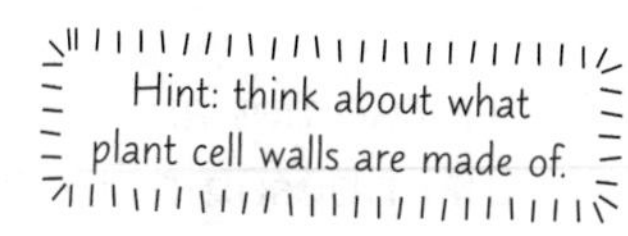
Hint: think about what
plant cell walls are made of.

Cell biologists — they don't get out much...

...that's because they're kept in cells... behind bars... Oh never mind. Cells are pretty straightforward — perfect for easing you in gently to GCSE Combined Science. That's no excuse for not having a go at all the questions on these two pages.

Microscopy

If I had to imagine a biologist's online dating profile, it would read something like this: 'Likes — long walks, sunsets, peanut butter, playing about with microscopes'.

Warm-Up

Microscopy means using a microscope to look at things.

Circle the correct word in each pair to complete the sentences below.

Microscopes magnify things. This means they make things look **bigger** / **smaller**.

Electron microscopes have a **higher** / **lower** magnification than light microscopes.

Electron microscopes have a higher resolution than light microscopes.
This means they allow things to be seen in **more** / **less** detail.

Q1 Which of the following would you need an electron microscope to see?
Tick **two** correct boxes.

- ☐ **A** a flea
- ☐ **B** a cell from an onion skin
- ☐ **C** the nucleus of a cell from an onion skin
- ☐ **D** the membranes inside a chloroplast
- ☐ **E** the plasmids inside a bacterial cell

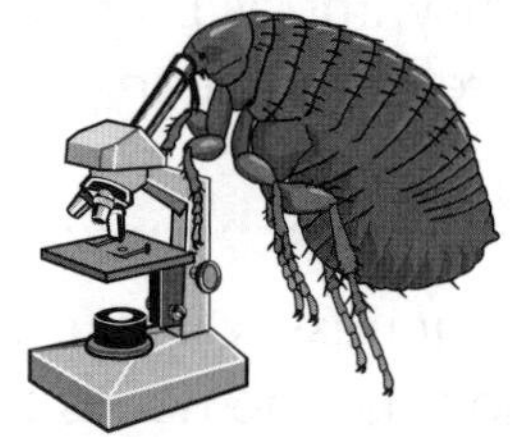

PRACTICAL

Q2 Padma wants to look at cheek cells using a light microscope. Cheek cells have no colour. Suggest why Padma adds a drop of methylene blue to the cells when she prepares her slide.

....

....

Q3 Ribosomes are tiny subcellular structures. Suggest why scientists didn't know that ribosomes existed until after the electron microscope was invented.

....

....

....

....

Q4 Dan looked at some cork cells under the microscope. Dan's teacher asked him to do a scientific drawing of what he could see. Dan's drawing is shown in **Figure 1**.

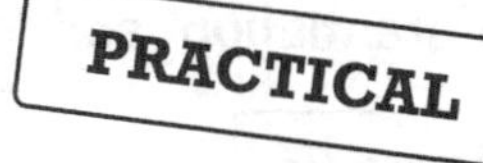

PRACTICAL

Figure 1

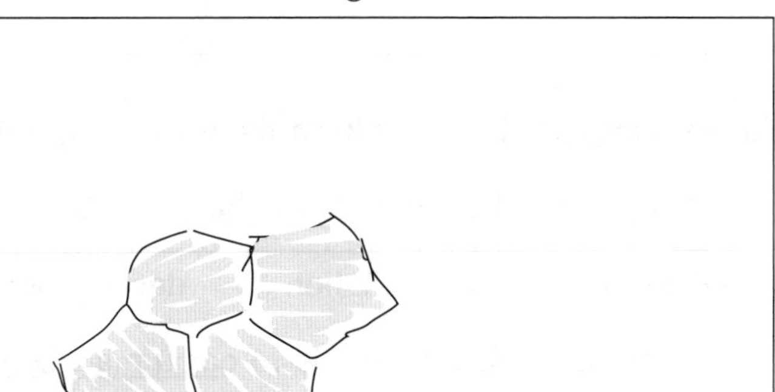

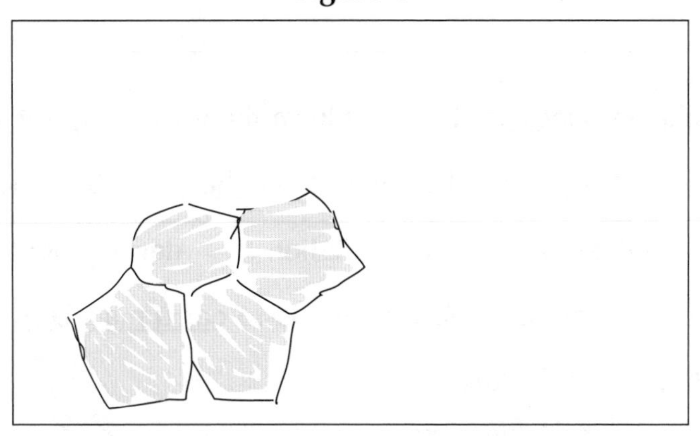

What could Dan do to make his scientific drawing better?
Tick **one** box.

☐ **A** Make his colouring in neater.
☐ **B** Include some of the cell details that he can't see under the microscope.
☐ **C** Draw the outlines of the cells using smooth lines.
☐ **D** Make his drawing smaller.

Q5 Professor Smart has invented a shrinking ray. She uses it to shrink her pet cat, Fluffy. Fluffy is now 0.02 mm long.

a) The Professor looks at Fluffy using a light microscope.
The image that she sees is 8 mm long.

i) Use the formula below to calculate the magnification of Professor Smart's image.

$$\text{magnification} = \frac{\text{image size}}{\text{real size}}$$

magnification = × ....

ii) What is the length of the image Professor Smart sees in micrometres (μm)?

1 millimetre (mm) = 1000 micrometres (μm)

.... μm

b) The Professor now looks at Fluffy using a × 600 magnification.
Use the formula below to calculate how long Fluffy will be in the new image.

image size = magnification × real size

.... mm

Biologists are big kids — they're always messing about with slides...

Make sure you practise using the magnification formula. You need to be able to apply it to question contexts you've not seen before (even silly ones about cats). That way, if there's ever a question where you have to use it, you'll know exactly what you need to do. Make sure you know how to convert between different units too, e.g. mm and μm.

Cell Differentiation and Specialisation

'Differentiation' and 'specialisation' are two long words to describe something fairly simple...

Warm-Up

Most cells in the body are <u>specialised</u>. This allows them to carry out <u>specific functions</u>.

<u>Circle</u> the name given to the process by which cells <u>change</u> to become specialised.

diffusion   differing   differentiation   development

<u>Sperm cells</u> are specialised cells. Their function is to <u>carry male DNA</u> to an <u>egg</u> cell.

Label this sperm cell with <u>one</u> feature that makes it specialised for its function.

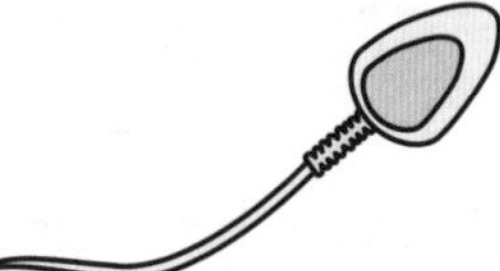

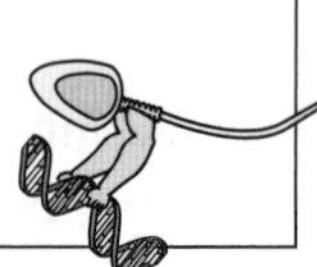

Q1 **Table 1** shows some specialised cell types. Fill in the gaps to complete the table.

Table 1

Cell Type	Function	Feature that helps to make cell specialised for its function
	Transports water around a plant.	Forms hollow tubes.
	....	Not many subcellular structures.
	....	Long, with branches at the ends.
	Contraction, which brings about movement.	....

Q2 Liver cells are a type of specialised cell. They carry out lots of chemical reactions, which need a large amount of energy. Suggest why liver cells have lots of mitochondria.

....

....

My specialisation is avoiding work...

The features I have that allow me to do this include having a Netflix account and being easily distracted. Don't you get distracted from the business of learning about specialised cell types and how they're adapted to carry out their functions.

Chromosomes and Mitosis

Since you were born, most cells in your body have been replaced several times as part of the cell cycle.

Warm-Up

A cell nucleus contains genetic information in the form of chromosomes.
Tick the correct box to complete each of the following sentences.

Chromosomes are made of...

...glucose ☐ ...DNA ☐ ...RNA ☐

In human body cells, chromosomes are found in...

...pairs ☐ ...triplets ☐

How did you say you wanted the cell dividing?

Mitosis is the part of the cell cycle in which cells divide.

Before a cell divides by mitosis, it has to...

...halve its DNA ☐ ...double its DNA ☐

Q1 Only **one** of the following statements about human chromosomes is **true**. Tick the correct one.

☐ **A** There are two chromosome 7s in a human nucleus.
☐ **B** There are three chromosome 7s in a human nucleus.
☐ **C** There is only one chromosome 7 in a human nucleus.
☐ **D** There are only seven chromosomes in a human nucleus.

Q2 A horse cell nucleus has 64 chromosomes. A chicken cell nucleus has 78 chromosomes. Chromosomes are found in pairs in both horse and chicken cells. How many more **pairs** of chromosomes does a chicken have than a horse?

.... pairs of chromosomes

Q3 Which of the following is a situation in which humans would **not** use mitosis to produce new cells? Circle the correct answer.

As an embryo, to develop in the womb.

As an adult, to produce cells that are genetically different for reproduction.

As a child, to produce new cells and increase in height.

As an adult, to produce new muscle cells and develop bigger muscles.

Q4 The following diagrams show the different stages of the cell cycle. Draw lines to match the description of each stage with the correct diagram.

a)

b)

c)

d)

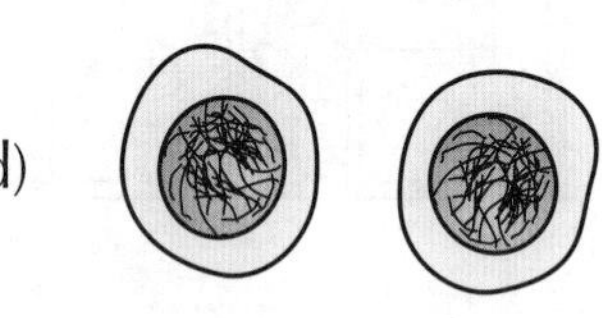

A membrane forms around each set of chromosomes. These become the nuclei for the new cells.

The cytoplasm and cell membrane divide. Two genetically identical cells are made.

The chromosomes line up across the centre of the cell. The arms of each chromosome are pulled to opposite ends of the cell.

Before a cell divides, it copies (replicates) its DNA. X-shaped chromosomes form.

Q5 Lucy had a small cut on her hand. A few days later she noticed that the cut had disappeared. The new skin looked just the same as the skin on the rest of her hand.

a) Lucy produced new skin cells by mitosis.
Suggest why this means that the new skin looks the same as the skin before it got cut.

....

....

b) Lucy's skin cells spend around 20% of the cell cycle in mitosis.

i) If the cell cycle for skin cells takes 24 hours, how many hours do the skin cells spend in mitosis?

.... hours

ii) Suggest **two** things that might be happening in skin cells that are not undergoing mitosis.

1. ....

2. ....

Students spend around 20% of their lesson time staring out the window...

Mitosis is one of those technical terms that sounds much scarier than it really is. It's just a type of cell division that makes two genetically identical cells. Simple really. Make sure you've got mitosis straight in your head because there's another type of cell division (with a very similar name) that you'll learn about later on. Right, on to the next page...

Stem Cells

Watch yourself. This topic contains some cutting edge science...

Warm-Up

Cells differentiate (change) to become specialised for their job.
All multicellular organisms contain stem cells.

Stem cells are undifferentiated cells that can divide to produce more undifferentiated cells. These cells can differentiate into different types of specialised cell.

Sells stems

Draw lines to match each type of stem cell on the left to its description on the right.

Type of stem cell	Description
Embryonic stem cell	Can divide to produce cells that can develop into certain types of specialised cells only.
Adult stem cell	Can divide to produce cells that can develop into any type of specialised cell.

Write down one place in the human body where adult stem cells are found.

....

Write down one type of specialised cell that adult stem cells produce.

....

Q1 Norbert takes a cutting from his best houseplant. A cutting is a small part of a plant that has been cut off it, e.g. a stem with a few leaves. If a cutting is planted in some soil, it can grow into a whole new plant.

a) Cuttings must contain meristems.

i) What type of cells are found in meristems?

....

ii) Suggest how these cells help cuttings grow into a whole new plant.

....

....

b) Norbert's houseplant has red flowers. What colour would you expect the flowers to be on the plant Norbert grows from the cutting?

....

c) Norbert took the cutting because it's cheaper than buying a new plant. Suggest **one** more reason why Norbert may have wanted to take a cutting rather than buy a new houseplant.

....

....

....

Q2 A man has damaged the nerve cells in his spinal cord. The nerve cells carry information from the man's brain to the rest of his body. The damage means that the man can't walk. Scientists are investigating using stem cells to help repair the damage.

a) Tick the boxes to show whether the following statements about how stem cells could be used to treat the man are **true** or **false**.

	True	**False**
i) Stem cells could be used instead of nerve cells in the man's spinal cord.	☐	☐
ii) Stem cells could be used to produce new nerve cells, which would replace the damaged cells in the man's spinal cord.	☐	☐
iii) The stem cells could be used to stick the nerve cells in the man's spinal cord back together.	☐	☐

b) i) What type of stem cells would be the most useful for the man's treatment? Tick **one** box.

☐ **embryonic stem cells** ☐ **adult stem cells**

ii) Explain your answer to part **b) i)**.

....

....

iii) Why might some people be against scientists using this type of stem cell to treat the man?

....

....

c) Why might scientists test the stem cells used in medical treatments to see if they contain viruses?

....

....

Meristems — the happiest parts of a plant...

Stem cell treatments have the potential to transform medicine, but we're not there yet. There are ethical issues to navigate along the way too. Make sure you know both the pros and cons of using stem cells in medicine.

Diffusion

Of all the different types of 'fusion', I think 'diffusion' is my favourite...

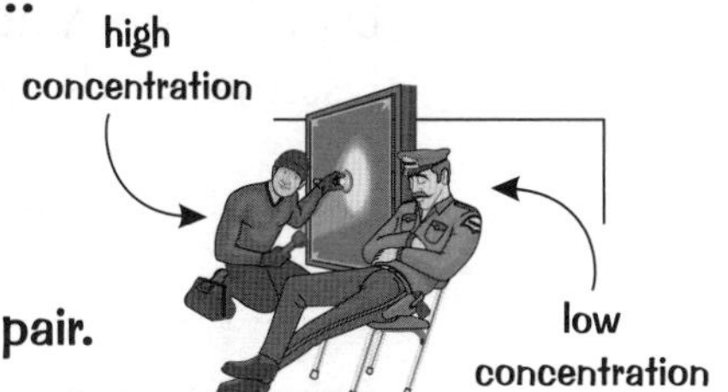

Warm-Up

Diffusion is how particles <u>move around</u> and <u>spread out</u>.

Complete the passage below by circling the correct word in each pair.

Diffusion is the movement of particles from an area where they are at a **higher** / **lower** concentration to an area where they are at a **higher** / **lower** concentration.

Q1 **Figure 1** shows a cup of hot water which has just had a drop of orange squash added. **Figure 2** shows the same cup a few minutes later.

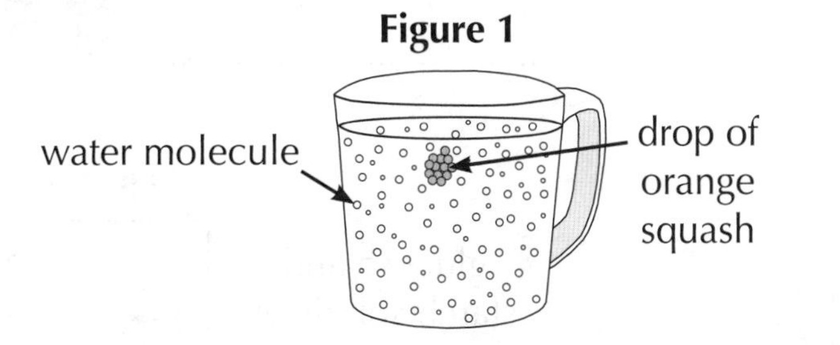

Figure 1

water molecule

drop of orange squash

Figure 2

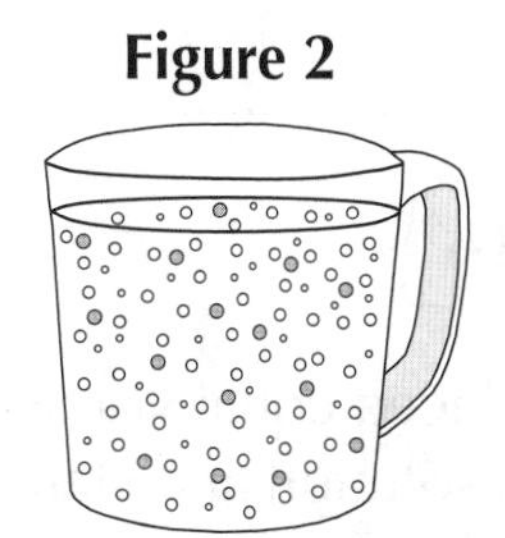

a) The passage below describes what's happened to the squash particles. Complete the passage using the correct words from the box.

water	lower	drop of squash	higher

The squash particles have moved from an area of .... concentration (the ....) to an area of .... concentration (the ....).

b) The cup is refilled with cold water and the same volume of squash is added. Do you think the rate of diffusion would be faster, slower or the same?

....

Q2 **Figure 3** shows two models of diffusion.

Figure 3

A

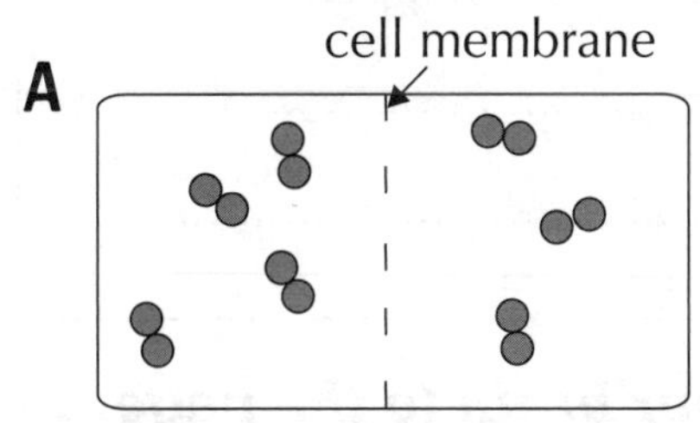

B

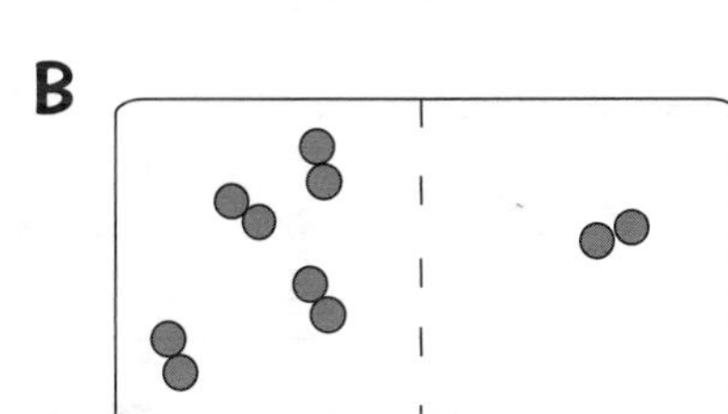

Would you expect the molecules to diffuse faster in model **A** or **B**? Why?

....

Q3 Madison was studying in her bedroom. Her dad was cooking curry for tea in the kitchen. Soon Madison could smell the curry that her dad was making.

a) Explain how Madison was able to smell the curry from the bedroom. Use the words 'curry particles', 'diffuse' and 'concentration' in your answer.

....

....

b) After tasting the curry, Madison's dad added more curry powder. The smell of curry got stronger. Suggest why.

....

....

c) Madison can smell the curry more strongly when it is hot than later on when it is cold. Suggest why.

....

....

PRACTICAL

Q4 Arjun placed equal volumes of glucose solution and starch solution inside a bag designed to act like a cell membrane. He then put the bag into a beaker of water (see **Figure 4**).

Figure 4

glucose and starch solutions

'cell membrane' bag

water

a) After 20 minutes, Arjun tested the water outside of the bag. He found that there was glucose in it. Explain how the glucose got into the water.

....

....

b) Starch is a big molecule. Arjun didn't find any starch in the water outside the bag. Suggest why.

....

c) Arjun did the experiment again using the same volumes of glucose and starch solutions. This time he used a much longer, thinner bag (see **Figure 5**).

Arjun found that the glucose moved into the water more quickly. Explain why the longer, thinner bag caused this to happen.

Figure 5

....

....

If only the answers could diffuse from your brain to the page...

Biology-wise, diffusion is what I'd call a key concept. It comes up again and again, often where you least expect it. So, trust me when I say that the doing the questions on these two pages is time well spent.

Osmosis

Think of it as a computer game — you've completed level 1 (diffusion), now it's time for level 2...

Warm-Up

Osmosis involves the movement of water molecules across a partially permeable membrane.
What is a partially permeable membrane? Tick one box.

- [ ] A membrane that lets both big and small molecules pass through it.
- [ ] A membrane that only lets small molecules pass through it.
- [ ] A membrane that only lets water pass through it.
- [ ] A membrane that doesn't let anything pass through it.

Which of the following statements about the movement of water molecules is true?
Tick one box.

- [ ] All the water molecules move towards the more concentrated solution.
- [ ] All the water molecules move towards the less concentrated solution.
- [ ] Water molecules move both ways across the membrane, but the overall movement is to the more concentrated solution.
- [ ] Water molecules move both ways across the membrane, but the overall movement is to the less concentrated solution.

Q1 **Figure 1** shows a model cell in water.

Figure 1

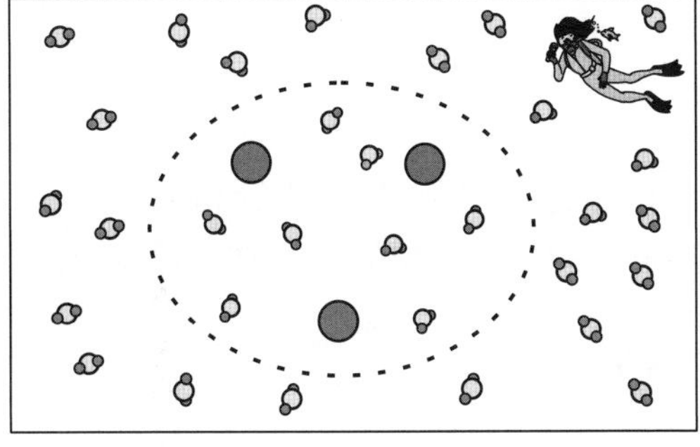

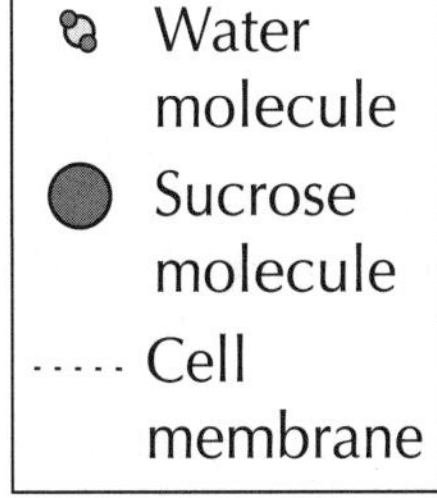

a) Would you expect water molecules to move **into** or **out of** the cell?

....

b) If the cell stays in the water, which of the following is most likely to happen?
Tick **one** box.

- [ ] **A** The cell will get bigger.
- [ ] **B** The cell will get smaller.
- [ ] **C** The cell will stay the same size.

PRACTICAL

Q2 Some parsnip chips were placed in solutions of different salt concentrations. At the start of the experiment each chip was 50 mm long. Their final lengths and percentage changes in length are recorded in **Table 1**.

Table 1

Concentration of salt solution (mol/dm^3)	Final length of parsnip chip (mm)	% change in length of parsnip chip
0.00	55	+10
0.25	53	
0.50	49	–2
0.75	51	+2
1.00	46	–8

a) i) By how many mm did the parsnip chip in the 0.25 mol/dm^3 salt solution change in length?

.... mm

ii) Calculate the percentage change in length for the parsnip chip in 0.25 mol/dm^3 salt solution.

................%

b) State the concentration of salt solution that produced an anomalous result.

An anomalous result doesn't fit the pattern of the other results.

.... mol/dm^3

c) Suggest a concentration of salt solution in which the change in the length of the parsnip chip would have been 0%.

.... mol/dm^3

Q3 Luiza was preparing some salted ham. She covered the ham in water and left it to soak for a few hours. When she returned, the ham was much bigger in size.

Use the term osmosis to help you explain the change in the ham.

....

....

....

Root vegetables — an a-peeling way to think about osmosis...

Osmosis is kind of like diffusion, but with water. And it has to happen across a partially permeable membrane. So really, they could just call it 'the diffusion of water through a membrane', but scientists love a snappy title.

Active Transport

Active transport is like cycling to the cinema instead of getting your dad to drive you. Sort of.

Warm-Up

Active transport is another way for substances to move **into** and **out of** cells.
It works in the **opposite** way to **diffusion**.

Place a tick in the correct boxes to show the features of each process in the table.

Feature	Diffusion	Active Transport
Substances move from areas of higher concentration to areas of lower concentration		
Requires energy		

Describe one way in which humans use active transport.

....

....

Q1 **Figure 1** shows a specialised plant cell.

Figure 1

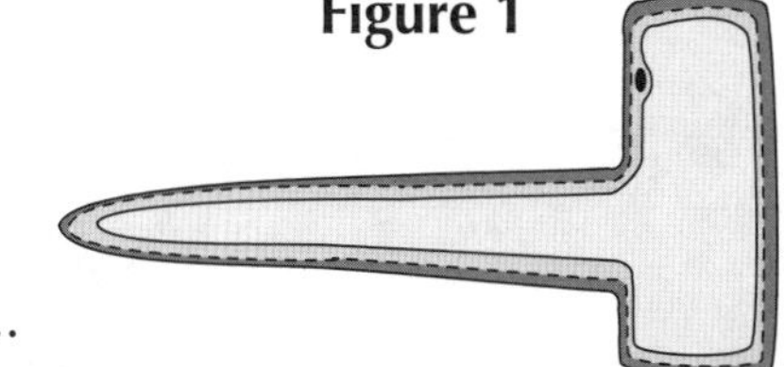

a) Name the type of cell shown.

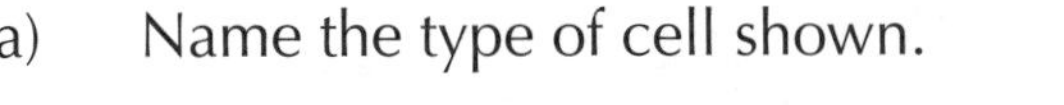

....

b) The passage below explains how these specialised cells absorb mineral ions from the soil. Use words from the box to fill in the gaps.
You **don't** need to use all of the words.

energy	lower	active transport	higher	respiration

The concentration of mineral ions in the soil is .... than in the root hair cells. The cells use .... to move the mineral ions into the root hair cells. This requires .... from ....

c) **Figure 2** shows a shorter version of the cell in **Figure 1**.
Suggest why this cell won't be able to absorb mineral ions from the soil as quickly as the cell in **Figure 1**.

Figure 2

....

Is that a plant cell or a tent peg?

Phew. All that active learning is tiring, but you're very nearly done with the fascinating subject of 'ways in which things move into and out of cells'. Bet you never thought there was this much to know. Just two more pages to go...

Exchanging Substances

Organisms have to get some stuff into them and other stuff out. Bit like the hokey cokey...

Warm-Up

Organisms need to <u>exchange substances</u> with their environment.
<u>Multicellular organisms</u> have <u>specialised exchange surfaces</u> in order to do this.
These exchange surfaces are <u>adapted</u> for efficient exchange.

For example, the surface of the <u>small intestine</u> is covered with <u>villi</u>.

Give <u>two</u> ways in which the villi make the small intestine well-adapted for absorbing nutrients.

1. ....

2. ....

Q1 In humans, gas exchange happens in the alveoli. **Figure 1** shows some healthy alveoli and some alveoli that have been damaged by smoking.

Figure 1

healthy alveoli

alveoli damaged by smoking

a) The alveoli that have been damaged by smoking have a smaller surface area than the healthy alveoli.

How will this affect the rate at which oxygen diffuses from the alveoli into the blood? Tick **one** box.

☐ **A** The rate of oxygen diffusion will increase.

☐ **B** The rate of oxygen diffusion will decrease.

☐ **C** The rate of oxygen diffusion will not be affected.

Figure 2 shows some alveoli with walls that have been made thicker through scarring.

Figure 2

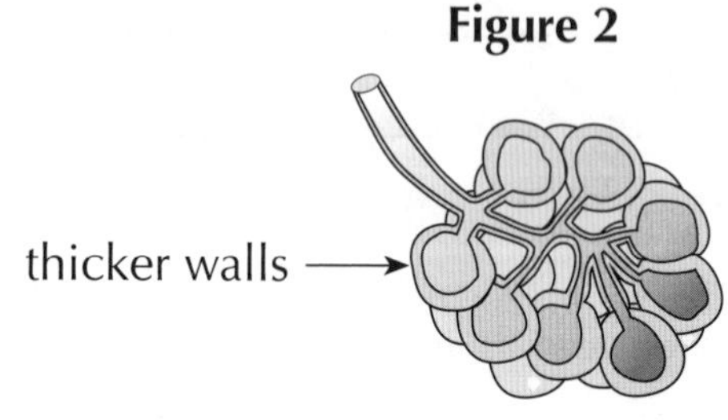

b) Suggest how the scarring will affect the rate at which oxygen diffuses from the alveoli into the blood. Explain your answer.

The rate of oxygen diffusion will .... **because** ....

....

c) In fish, gas exchange happens in the gills. Like alveoli, gills are adapted for gas exchange. Write down **two** similarities in the way gills and alveoli are adapted for gas exchange.

1. ....

2. ....

PRACTICAL

Q2 Kwame did an experiment to investigate the effect of surface area to volume ratio on the rate of diffusion. He used four agar cubes of different sizes.

The cubes contained a dilute alkaline solution and a pH indicator that made them pink. Kwame placed the cubes in a beaker of dilute acid and measured how long it took for the acid to diffuse into the centre of the cubes and change the colour of the pH indicator from pink to colourless. His results are shown in **Table 1**.

Table 1

Dimensions of cube (cm)	Surface area to volume ratio	Time taken for cubes to change colour (s)
1 × 1 × 1	6 : 1	186
2 × 2 × 2	3 : 1	325
5 × 5 × 5	1.2 : 1	542
10 × 10 × 10	0.6 : 1	745

a) Complete these statements by circling the correct word in each pair.

i) As the cubes get bigger, their surface area to volume ratio becomes **bigger** / **smaller**.

ii) As the surface area to volume ratio gets smaller, the time taken for the cubes to change colour **increases** / **decreases**.

iii) The longer the cubes take to change colour, the **more** / **less** time it takes for the acid to diffuse into the centre of the cubes.

iv) So the smaller the surface area to volume ratio, the **slower** / **faster** the rate of diffusion into the centre of the cubes.

b) Multicellular organisms have a small surface area to volume ratio. Using the results of Kwame's experiment, suggest why multicellular organisms are unable to exchange everything they need to survive across their outer surface.

....

....

If you're bored, work out the surface area : volume ratio of a loved one...

Exchange surfaces are all adapted to speed up the rate of diffusion — so factors that affect the rate of diffusion (e.g. surface area, concentration gradient) will influence the adaptations that exchange surfaces have. Right, that's Topic B1 — Cell Biology over and done with. Reward yourself with 10 minutes of cat videos and a cup of tea.

Cell Organisation

Congratulations, you've made it to Topic B2. Your prize is a few questions on cell organisation.

Warm-Up

Multicellular organisms contain lots of cells.
These cells are grouped into different levels of organisation.

Complete the passage below using words from the box.

organ | cells | organ system | function | organisms

A tissue is a group of similar .... that work together to perform a particular .... Different tissues that work together make up an .... An .... is formed by several different organs working together.

Q1 Samina is looking at a layer of human skin under a microscope. This layer of skin acts as a physical barrier, which helps to protect the body from the environment. Samina makes a simple drawing of what she can see. The drawing is shown in **Figure 1**.

Figure 1

Cells

a) i) What level of organisation is shown in **Figure 1**? Circle the correct answer.

Organ system | Organ | Tissue | Organism

ii) Explain your answer to part **a) i)**.

....

....

b) Samina looks at some different layers of skin under her microscope. The layers work together to protect the body. She notices that each layer is made up of a different type of cell. Based on her observations, Samina thinks skin is an organ.
Is she correct? Explain your answer.

....

....

Q2 A human organ system is shown in **Figure 2**.

Figure 2

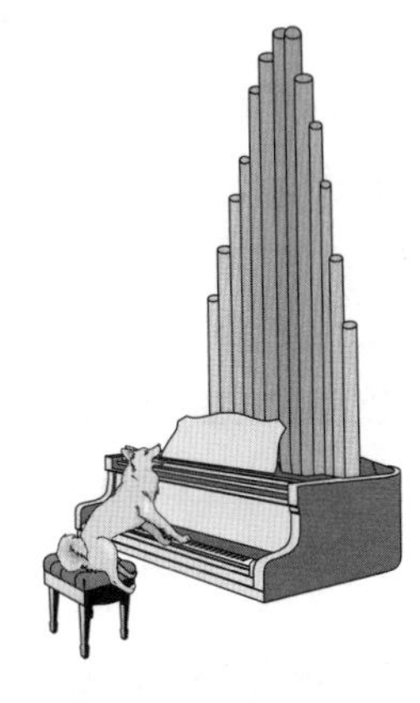

a) i) What is the name of the organ system shown in **Figure 2**?

....

ii) Which letter on the diagram (**A**, **B** or **C**) represents the small intestine? ....

b) Why is the small intestine described as an organ?
Tick **one** box.

☐ **A** It is a group of similar cells that work together to carry out a function.

☐ **B** It is a group of different tissues that work together to carry out a function.

☐ **C** It is a group of identical tissues that work together to carry out a function.

Q3 The circulatory system is an organ system that carries blood around the body. The structures listed in **Table 1** are all part of the circulatory system.

Write the numbers **1-5** in the table to show how the structures are organised, from smallest to largest. The first one has been done for you.

Table 1

Structure	Order
epithelial tissue	....
nucleus	1
heart	....
circulatory system	....
epithelial cell	....

I wish my desk was as well organised as my organ systems...

Inside each level of organisation in your body is another level of organisation. Just like Russian dolls. Sort of. Except Russian dolls don't have organs inside them, just smaller Russian dolls. So I guess maybe it's nothing like that after all.

 ☐  ☐  ☐

Enzymes and Their Reactions

The job of enzymes is to make the reactions in your body happen faster. But they can be pretty picky about the conditions that they'll work in. Scientists can run experiments to test this.

Warm-Up

Enzymes are catalysts.
A catalyst is a substance that increases the speed of a reaction.
Catalysts do not get changed or used up during the reaction.

Complete the sketch to show how the enzyme's shape lets it break down substances.

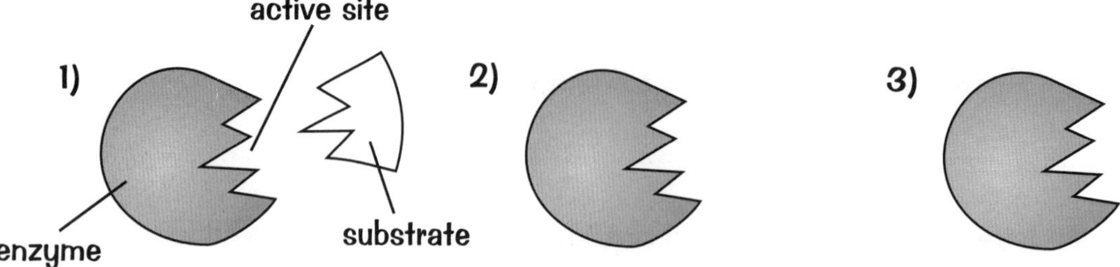

Q1 Keith has poured melted fat into his kitchen sink. The fat has blocked the sink's drain. Keith buys a bottle of drain cleaner to try to unblock it. The drain cleaner contains enzymes.

a) The drain cleaner breaks down 10 grams of fat in 200 seconds.
Calculate the rate of this reaction. Give your answer in grams per second (g/s).

rate = ... g/s

b) Suggest how the enzymes in the drain cleaner help to unblock the drain.

....

c) Keith pours boiling water into the sink five minutes after putting the drain cleaner in.
His drain remains blocked.

i) Suggest what effect the boiling water had on the enzymes' active sites.

....

....

ii) Suggest why the drain cleaner hasn't worked.

....

....

....

PRACTICAL

Q2 The optimum pH of an enzyme is the pH that the enzyme works best at. Elinah has a sample of enzyme **X** and she is trying to find its optimum pH. Elinah tests enzyme **X** by timing how long it takes to break down a substance at different pH levels. The results of Elinah's experiment are shown in **Table 1**.

Table 1

pH	Time taken for reaction to complete (seconds)
2	100
4	82
6	18
8	76
10	98
12	100

a) Plot the results from **Table 1** on the grid in **Figure 1**. Join the points with straight lines.

Figure 1

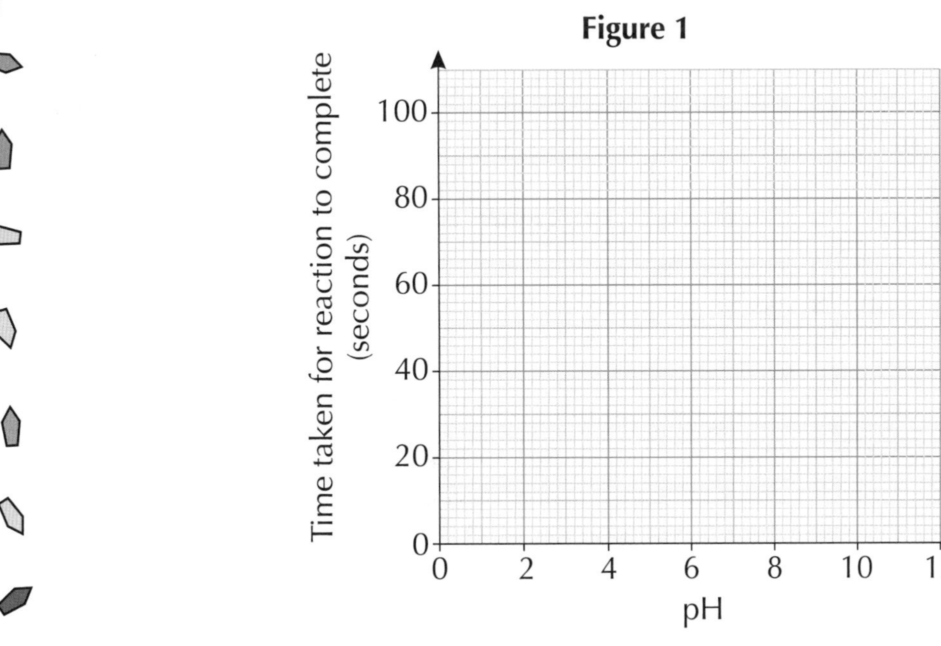

b) Roughly, what is the optimum pH for this enzyme? Tick **one** box.

☐ pH 2 ☐ pH 6 ☐ pH 8 ☐ pH 12

c) The pH in the stomach is roughly pH 2.
Would you expect to find enzyme **X** in the stomach? Explain your answer.

....

d) Describe **one** thing that Elinah should have done to make sure her experiment was a fair test.

....

I love enzyme experiments — they're pHantastic...

Enzymes really are pretty nifty — they do all sorts of jobs in all different types of organisms. Remember that the activity of enzymes is affected by both temperature and pH, and be prepared to interpret data about enzyme activity.

 ☐  ☐  ☐

Enzymes, Digestion and Food Tests

Different enzymes are responsible for breaking down different molecules in our food. We can find out what type of molecules are present in our food using food tests. What fun.

Warm-Up

Enzymes are released into the gut to break down substances in food.

Complete the sentences below by circling the correct word or words from each pair.

Carbohydrases break down **lipids / carbohydrates** into sugars.

Amylase is a carbohydrase that breaks down **starch / lactose**.

Amylase is made in the **stomach / salivary glands**.

Circle the name of the test below that can be used to detect starch in a food sample.

iodine test **Benedict's test** **Sudan III test**

Q1 Two different species of bacteria were transferred to an agar plate. The agar contained a lipid which made it cloudy. Agar that doesn't contain a lipid is usually clear. The plate was then left overnight. **Figure 1** shows the results.

a) What type of enzyme breaks down lipids?

....

b) Which species of bacteria in the diagram (**A** or **B**) contains this enzyme?

....

Figure 1

Cloudy agar

Clear area of agar

Bacterial species **A**

Bacterial species **B**

Q2 **Figure 2** shows part of the human digestive system. Tube **X** transports bile from where it is produced to the gall bladder. Tube **Y** transports bile from the gall bladder to the small intestine. Tube **Y** can sometimes get blocked.

Figure 2

Gall bladder

Tube **X**

Tube **Y**

Small intestine

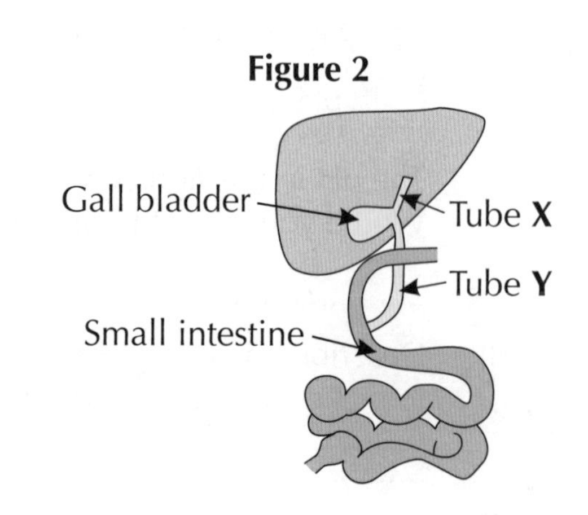

a) Which organ does tube **X** transport bile from?

....

b) Which of following is likely to happen if tube **Y** becomes blocked? Tick **one** box.

☐ **A** More fat will get emulsified in the small intestine.

☐ **B** The small intestine will become more acidic.

☐ **C** The rate of fat breakdown in the small intestine will increase.

Q3 Martha is cooking some beef for dinner. The cut of beef is quite tough because it contains a lot of collagen (a protein). Martha soaks the beef in some pineapple juice for 30 minutes before cooking it. This helps to make the beef softer and easier to chew.

a) Pineapple juice contains a type of protease enzyme. What do proteases break down?

....

b) Suggest how the pineapple juice makes the beef easier to chew.

....

....

....

PRACTICAL

Q4 Two scientists are preparing to do an experiment in the lab, but they forget to label their beakers. They have four beakers (**A**, **B**, **C** and **D**). Each one contains either starch, a sugar solution, proteins or pure water. The scientists carry out several tests to determine what is in each beaker. **Table 1** shows the results of all the different tests that the scientists carried out on the contents of the beakers.

Table 1

	A	B	C	D
Biuret test	Purple	Blue	Blue	Blue
Iodine test	Browny-orange	Browny-orange	Browny-orange	Blue-black
Benedict's test	Blue	Blue	Brick-red	Blue

a) Based on the results in **Table 1**, write down the letter of the beaker (**A-D**) that contains each substance.

Beaker .... contains starch.

Beaker .... contains a sugar solution.

Beaker .... contains proteins.

Beaker .... contains pure water.

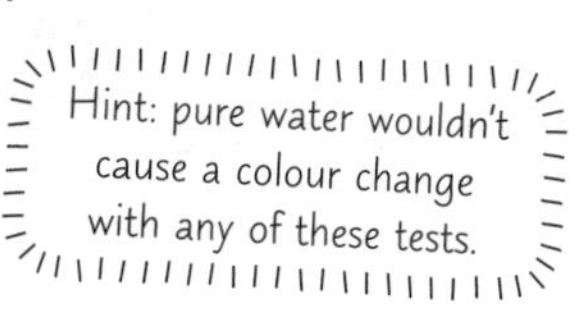

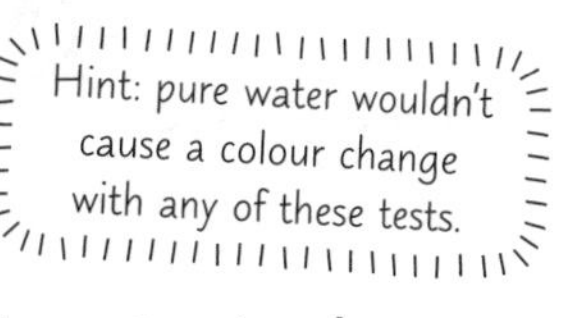

b) For which of the tests in **Table 1** would the scientists have had to heat the contents of the beakers to 75 °C?

....

My car breaks down so much I decided to call it Enzyme...

To me, science is all about adding chemicals to things and watching them change colour. So you can imagine how much fun I have doing food tests. Make sure you know how to carry out food tests, as well as how to interpret the results.

The Heart and Lungs

Now for something a little different — two very important organs. The heart and lungs work closely together to get oxygen all the way around your body. What a team.

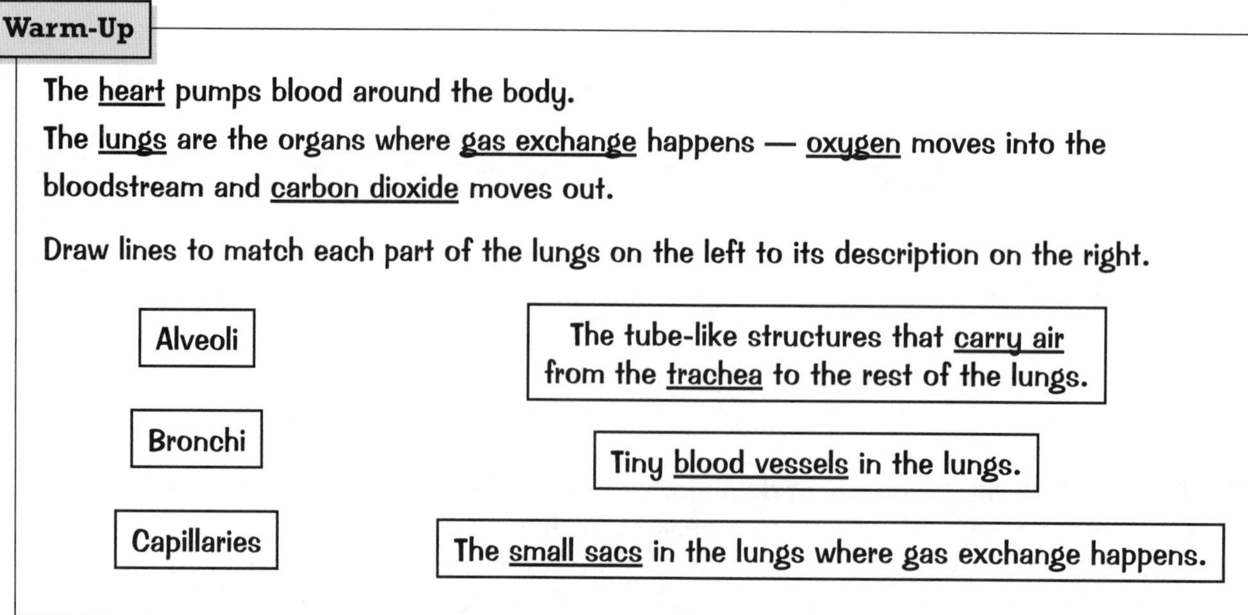

Warm-Up

The heart pumps blood around the body.
The lungs are the organs where gas exchange happens — oxygen moves into the bloodstream and carbon dioxide moves out.

Draw lines to match each part of the lungs on the left to its description on the right.

Alveoli	The tube-like structures that carry air from the trachea to the rest of the lungs.
Bronchi	Tiny blood vessels in the lungs.
Capillaries	The small sacs in the lungs where gas exchange happens.

Q1 **Figure 1** shows the main blood vessels of the heart. The arrows show the direction of blood flow.

Figure 1

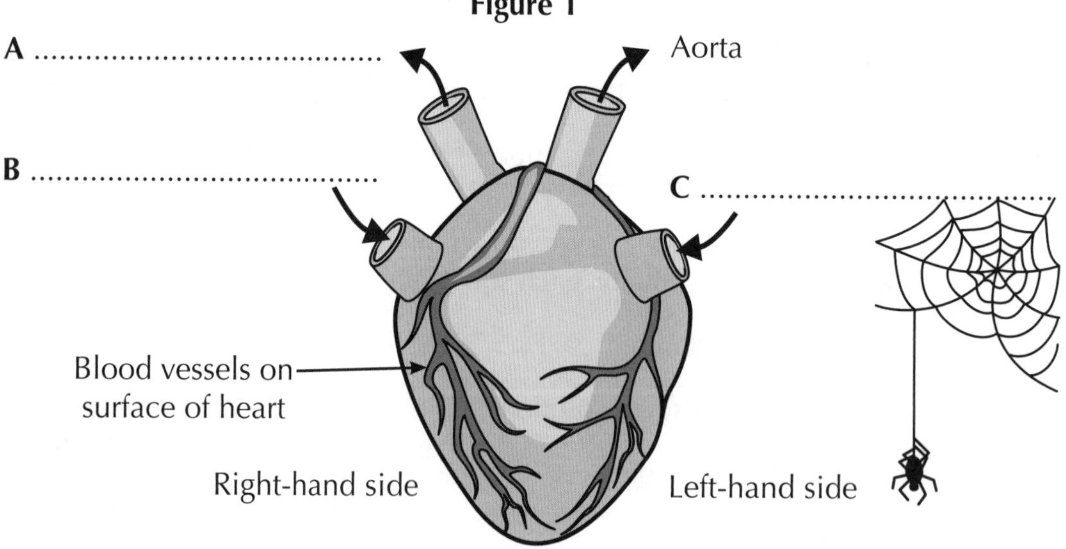

a) Write the names of the following blood vessels beside the letters on **Figure 1**.

pulmonary vein **vena cava** **pulmonary artery**

b) i) The blood carried by blood vessel **A** is... ☐ **A** oxygenated ☐ **B** deoxygenated

ii) Where does blood vessel **A** carry the blood to?

....

c) What are the blood vessels shown on the surface of the heart called?

....

Q2 **Figure 2** shows part of the respiratory system.

a) On **Figure 2**, name and label the tubes that branch off the trachea.

b) On **Figure 2**, label an alveolus.

Figure 2

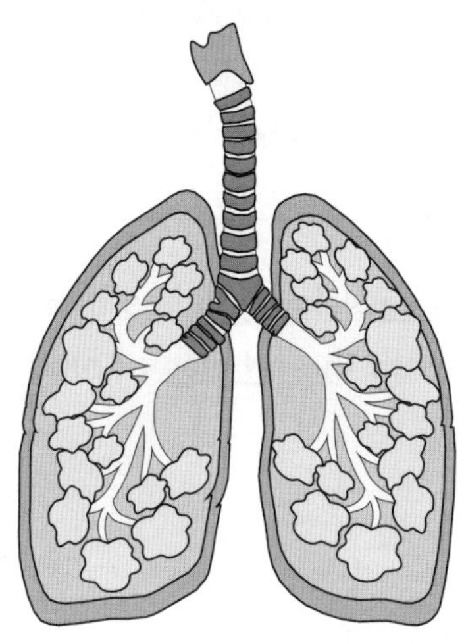

Q3 Hugh has an irregular heartbeat. An irregular heartbeat is when the heart doesn't beat as normal.

a) i) A doctor uses a heart rate monitor to measure Hugh's heart rate. The monitor records 140 beats in 240 seconds. Calculate Hugh's heart rate using the equation below.

Watch out: you need to convert 240 seconds into minutes here.

$$\text{heart rate} = \frac{\text{number of heart beats}}{\text{number of minutes}}$$

.... beats per minute

ii) Bradycardia and tachycardia are types of irregular heartbeat. **Table 1** shows when someone is considered to have bradycardia, tachycardia or a normal heart rate.

Table 1

Diagnosis	Heart rate (beats per minute)
Bradycardia	Below 60
Normal	60 - 100
Tachycardia	Above 100

Using this information and your answer to part **a) i)**, what do you think the doctor's diagnosis will be for Hugh?

....

b) Suggest a suitable treatment for Hugh. Explain your answer.

....

....

Lung-es are the best exercises for getting through this page...

The way that the lungs and heart work is a little tricky, so take your time to ensure you've got it all sussed. Make sure you learn the names of the blood vessels of the heart, and the order in which blood travels through them.

Circulatory System — Blood Vessels

Just in case you haven't had enough of the circulatory system, here's some more for you.

Warm-Up

Blood vessels are tubes that carry blood around the body.

There are three types of blood vessels. Use the words below to complete the definitions.

low	away from	towards
high	permeable	thin

Arteries carry blood .... the heart under .... pressure.

Veins carry blood .... the heart under .... pressure.

Capillaries have walls that are .... and ....

Q1 **Figure 1** shows how blood pressure changes as blood flows around the body.

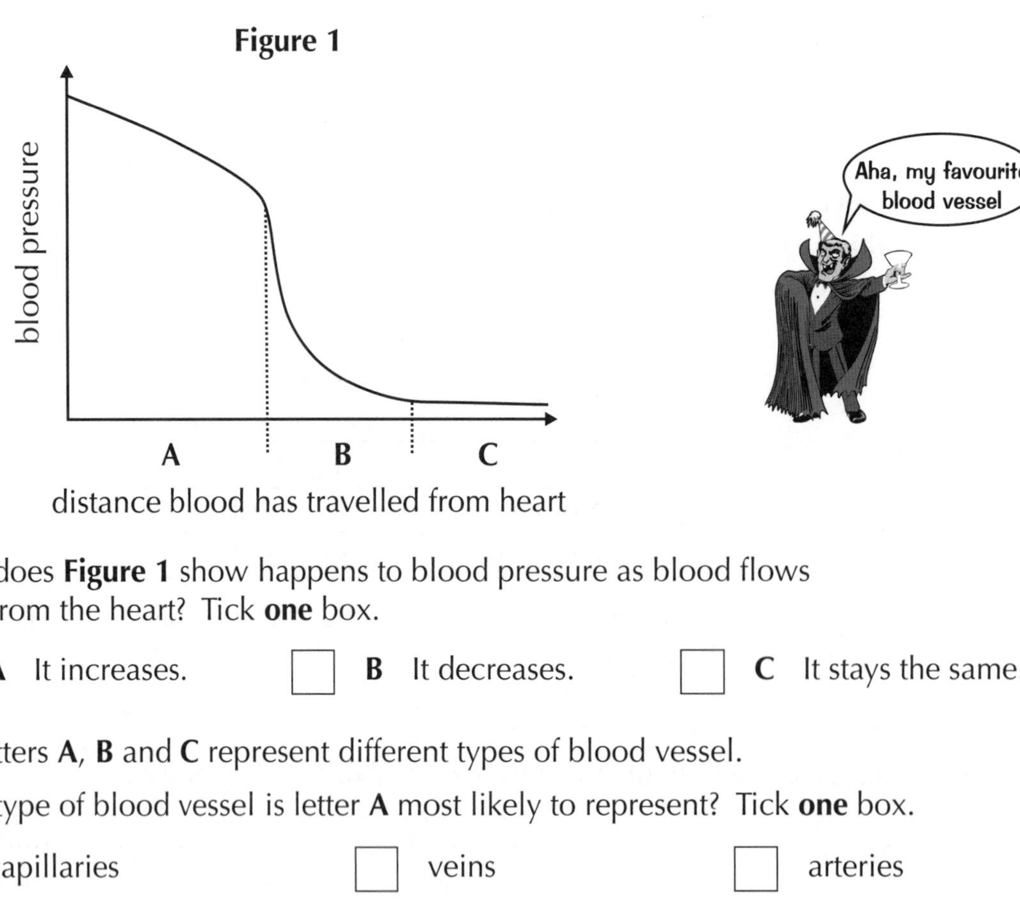

a) What does **Figure 1** show happens to blood pressure as blood flows away from the heart? Tick **one** box.

☐ **A** It increases. ☐ **B** It decreases. ☐ **C** It stays the same.

b) The letters **A**, **B** and **C** represent different types of blood vessel.

i) What type of blood vessel is letter **A** most likely to represent? Tick **one** box.

☐ capillaries ☐ veins ☐ arteries

ii) The blood vessels represented by **A** have more muscular walls than the blood vessels represented by **C**. Using the information in **Figure 1**, suggest why.

....

....

Q2 The blood flow through an artery in Brunhilda's leg was measured for three minutes whilst she was sitting down. In three minutes, 0.9 dm^3 of blood passed through the artery.

a) Calculate the rate of blood flow through the artery in Brunhilda's leg. Use the equation below. Give your answer in dm^3/min.

$$\text{rate of blood flow} = \frac{\text{volume of blood}}{\text{time}}$$

rate of blood flow = ... dm^3/min

b) Now calculate the rate of blood flow through the artery in Brunhilda's leg in cm^3/min.

1 dm^3 = 1000 cm^3

rate of blood flow = ... cm^3/min

Q3 Maya dissects two blood vessels from a piece of meat. One is a vein and the other is an artery. Measurements taken from both vessels are shown in **Table 1**.

Table 1

	Thickness of wall (mm)	Diameter of lumen (mm)
Blood vessel **A**	1.2	3.9
Blood vessel **B**	0.5	5.2

The lumen is the hole down the middle of a blood vessel.

a) i) Maya thinks blood vessel **B** is the vein. Use the information in **Table 1** to suggest **one** reason why blood vessel **B** is likely to be the vein.

....

....

ii) Suggest **one** other thing Maya could look for in blood vessel **B** to confirm it's the vein.

....

b) When tested, one of the blood vessels was found to be much more elastic than the other. Which one would you expect this to be (**A** or **B**)? Explain your answer.

....

....

Arteries are great at exams — they work well under pressure...

Although the heart does the brute force of the work getting blood around your body, blood vessels are pretty important too. So make sure you know which type of blood vessel does what, and how their structures are related to their functions.

Circulatory System — Blood

Blood is a bit like a lot of tiny lorries, working to transport all sorts of things around your body.

Warm-Up

Blood has four main parts — red blood cells, white blood cells, platelets and plasma.

Each part has a different function, and different features that allow it to carry out this function.

Tick the boxes to show whether the statements apply to red blood cells or white blood cells.

	Red Blood Cells	White Blood Cells
Have a nucleus.	☐	☐
Carry oxygen around the body.	☐	☐
Help protect against infection.	☐	☐
Shape gives them a large surface area.	☐	☐

Plasma is the liquid that carries everything in blood. Circle three things below that are carried by plasma.

hormones | bits of food | carbon dioxide | urine | glucose

Q1 Whales are marine mammals. They have lungs, so they can't breathe underwater. Instead, they have to go to the surface to breathe air. Whales have a high concentration of haemoglobin in their red blood cells. When they dive, they are able to hold their breath for a long time.

a) The passage below explains the function of the haemoglobin in a whale's red blood cells. Complete the passage by circling the correct word or words in each pair.

> The haemoglobin binds with **carbon dioxide / oxygen**.
>
> This allows the whale's red blood cells to pick up oxygen in its **lungs / heart** and transport it to the whale's **tissues / veins**.

b) Suggest why the high concentration of haemoglobin in a whale's red blood cells allows the whale to dive for long periods.

Q2 Jakob is looking at a sample of blood under a microscope. **Figure 1** shows what he can see.

Figure 1

X

Y

a) Jakob thinks cell **X** is a red blood cell.
Give **one** way Jakob can tell that cell **X** is a red blood cell from **Figure 1**.

b) What is cell **Y**?

Q3 Bernard-Soulier syndrome (BSS) is a rare disease that stops platelets from working properly. For people with BSS, even a minor injury can cause them to lose a lot of blood. Suggest why BSS causes excessive blood loss.

Where do you weigh a whale? At a whale-weigh station...

Here's a mildly interesting fact: blood with lots of oxygen is bright red and blood without oxygen is a darker red. It's sometimes possible to see blood vessels through the skin, which look like the blood in them is blue. These are veins, and the blood is actually a dark red-brown colour. It just looks blue due to the way light is reflected by the skin.

Cardiovascular Disease

You need to know some of the treatments for cardiovascular disease, and when they're used.

Warm-Up

Cardiovascular disease means a disease that affects the heart or blood vessels.

Complete the sentences below by circling the correct word or words from each pair.

Too much cholesterol in the bloodstream can cause **fatty** / **protein** deposits to form inside the coronary arteries. This can **increase** / **reduce** blood flow to the heart muscle. This means less **oxygen** / **carbon dioxide** can get to the heart muscle. This can result in a heart attack.

Q1 Read the descriptions of the patients in **Table 1**.

a) For each patient choose a suitable treatment (**A** to **D**) from the list below and write down the reason for your choice.

A artificial heart transplant

B statins

C biological replacement valve

D stent

Table 1

	Recommended Treatment	Reason for recommendation
Annie has a high blood cholesterol level.	....	....
Valerie is a 40-year-old woman with heart failure. She may die soon if she is not treated.	....	....
Alastair has a blocked artery to the heart muscle.	....	....

b) Suggest a disadvantage of the treatment you recommended for **Annie**.

....

I'm pretty sure the Tin Man must have had an artificial heart...

Cardiovascular disease is no laughing matter. The good news is that the death rate from cardiovascular disease has fallen over the past 50 years. But the treatments for cardiovascular disease can have disadvantages, which you need to know.

Health and Disease

You can think of health and disease as being opposite sides of the same coin...

Warm-Up

Health is a state of physical and mental wellbeing.
Diseases are things that affect someone's health.

Draw lines to match each type of disease below to the correct definition.

Communicable diseases	Cannot be spread from person to person or between animals and people.
Non-communicable diseases	Can be spread from person to person or between animals and people.

Tick the correct boxes to show whether the statements about health are true or false.

	True	False
Being under lots of stress improves someone's health.	☐	☐
The amount of money someone has can affect their health.	☐	☐
Someone's physical health has no effect on their mental health.	☐	☐

Q1 A doctor sees the following three patients in her surgery.

a) Anton does a lot of sport and eats a varied diet. However, he feels very anxious when taking public transport. He will walk long distances to avoid getting the bus. Do you think Anton is healthy? Explain your answer.

....

....

b) Grace has cystic fibrosis. This is an inherited disease caused by a faulty gene. It can lead to breathing problems and lung infections. Is cystic fibrosis a communicable or non-communicable disease?

....

c) Ollie has asthma. He needs a flu vaccine. The vaccine will reduce the risk of Ollie getting the flu. Explain why someone with asthma should try to reduce their chance of getting an infection such as flu.

....

....

If you don't like these jokes, you might have a disease in your funny bone...

There's not much to joke about on this page. Make sure you know the difference between communicable and non-communicable diseases. Check that you also understand that diseases can affect mental as well as physical health.

Risk Factors for Disease and Cancer

Risk factors are not fun, but the more we know about them, the more we can do to reduce the risk that people will get particular diseases.

Warm-Up

Risk factors are things that increase the likelihood that a person will develop a particular disease.

Lots of different things can be risk factors for disease.
Underline all the things below that you think are risk factors for disease.

smoking	eating broccoli	cycling
reading comic books	sunbathing	drinking too much alcohol

Q1 **Figure 1** shows the percentage of adults in Great Britain who said they were smokers between 1974 and 2014. It also shows the percentage of adults who said they were ex-smokers.

Hint: make sure you look at the key on the graph.

a) What percentage of adults were smokers in 1990?

.... %

Figure 1

b) Describe how the percentage of **ex-smokers** changed between 1974 and 2014.

....

....

....

....

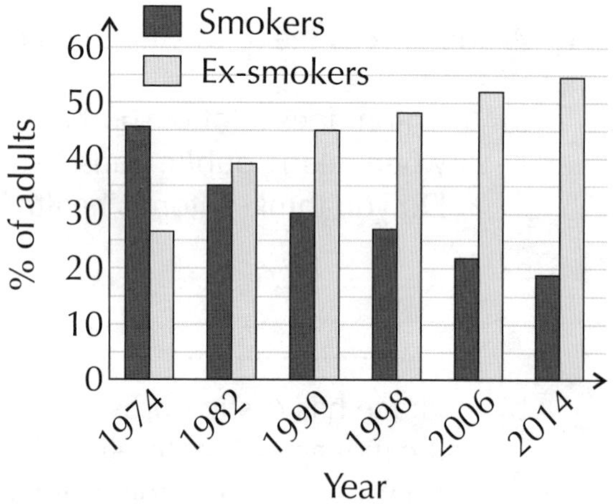

c) Why are smokers more likely to suffer from cancer than non-smokers? Tick **one** box.

☐ **A** Smoking exposes them to UV radiation.

☐ **B** Smoking reduces the level of cholesterol in their blood.

☐ **C** Smoking exposes them to carcinogens.

d) Describe **one** way in which cancer due to smoking may be expensive for the individual with cancer.

....

....

Q2 Some risk factors for disease are to do with a person's lifestyle.
Table 1 shows some information on the lifestyles of Tricia and Dave.

Table 1

Name	Fruit and vegetable portions per week	Cigarettes smoked per day	Hours of exercise per week
Tricia	10	0	0
Dave	35	40	10

a) i) Suggest **one** thing Tricia and Dave could each do to reduce their risk of developing cardiovascular disease.

Tricia: ....

Dave: ....

ii) Suggest **one** human cost of developing cardiovascular disease.

....

b) Based on the information in the table, which of the following diseases do you think Dave is at high risk of developing? Tick **one** box.

☐ **A** lung disease ☐ **B** measles ☐ **C** Type 2 diabetes

Q3 Frieda has been feeling unwell and has pain near her stomach. When she goes to the hospital, doctors find that she has a tumour in her liver. When they test the tumour, they find that the cells are from a tumour in Frieda's breast.

a) How do you think breast tumour cells were able to form a tumour in Freida's liver?

....

....

b) What type of tumour is the tumour in Freida's liver? Tick **two** boxes.

☐ benign ☐ malignant

☐ a primary tumour ☐ a secondary tumour

c) Freida's daughter wants to be tested to see if she has a faulty gene that causes cancer. Suggest why.

....

....

Writing bad jokes is a risk factor for your friends getting annoyed...

Risk factors for disease don't mean that someone will definitely get a disease, but they do make it more likely. Some risk factors (like smoking) can be avoided, but others (like genetics) are much trickier to dodge.

Plant Tissues, Transpiration & Translocation

That's enough of disease for the time being. Time to move on to the wonderful world of plants.

Warm-Up

Plant cells are organised into tissues and organs.
These each have different characteristics and functions.

Circle the correct word or words in each pair to complete the following sentences.

Palisade mesophyll tissue is the part of the leaf where most **water uptake / photosynthesis** happens. **Spongy / Rubbery** mesophyll tissue contains big air spaces so that gases can diffuse in and out of cells.

The loss of water from a plant is called **transpiration / translocation**.
Most water leaves the plant through the **waxy cuticle / stomata**.

Q1 **Figure 1** shows a simple diagram of a plant.

Figure 1

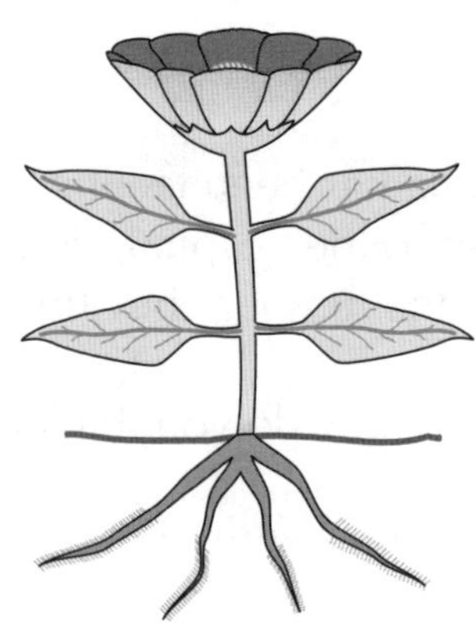

a) Put an **X** on the diagram to show one place where water enters the plant.

b) Add a **Y** to the diagram to show one place where water leaves the plant.

c) Add arrows to the diagram to show how water moves from the place where it enters to where it leaves.

Q2 Sweet potatoes grow from the roots of sweet potato plants, as shown in **Figure 2**. Sweet potatoes contain lots of starch. This starch is produced in the roots from sugar transported from other parts of the plant.

Figure 2

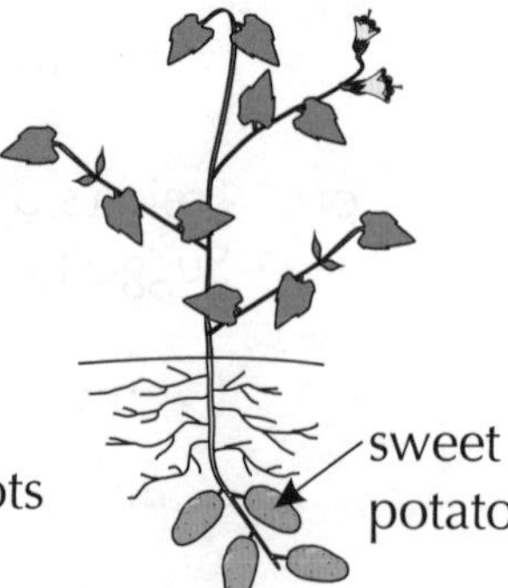

a) Which parts of a sweet potato plant produce most of the plant's sugar? Tick **one** box.

☐ **A** the flowers ☐ **B** the leaves ☐ **C** the roots

b) Which process transports sugar to the sweet potatoes? Tick **one** box.

☐ **A** translocation ☐ **B** transpiration ☐ **C** photosynthesis

Q3 Zahra has prepared flowers for a wedding. She has put a white carnation into a vase of water containing blue dye, as shown in **Figure 3**.

Figure 3

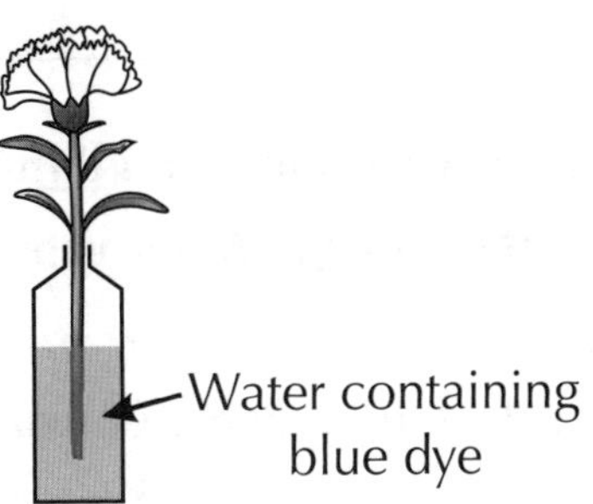

a) Which diagram (**A**-**D**) shows what the carnation and the vase will look like after a few days? Circle the correct letter.

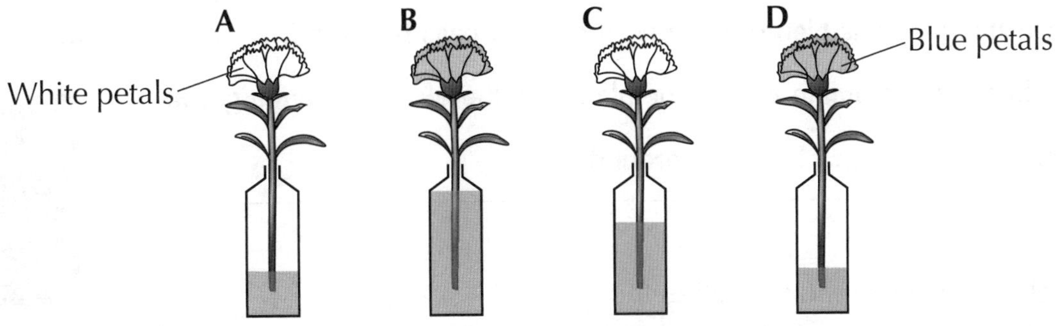

b) Explain your answer to part **a)**.

....

....

....

Q4 Nitrate is an example of a mineral ion. Plants need to take in nitrate from the soil to survive.

a) Which tissue is responsible for transporting nitrate in solution through plants?

....

b) Explain **one** way in which the structure of this tissue is adapted for its function.

....

....

....

Don't get stressed by plant biology — just relax and go with the phloem...

There are lots of words to learn when it comes to plants, but once you get your head around them hopefully you'll agree that plants are really cool. And maybe you'll even be inspired to play around with carnations and food colouring.

Transpiration and Stomata

Here are some more questions on transpiration. Bet you just can't wait.

Warm-Up

Transpiration is caused by evaporation and diffusion of water from a plant. Most water is lost from the leaves. Water vapour can escape from leaves through the stomata.

Complete the passage below using some of words from the box.

air	floppy	number	fat	size	soil

Guard cells are responsible for changing the .... of stomata.

When the plant has plenty of water, the guard cells become ....

This keeps the stomata open. The plant can therefore exchange gases with the .... around it.

Q1 Tick the boxes to show whether the following statements are **true** or **false**.

		True	False
a)	Transpiration rate is slower on a cloudy day than on a sunny day.	☐	☐
b)	Transpiration rate increases as it gets colder.	☐	☐
c)	The rate of transpiration is faster on a breezy day.	☐	☐

PRACTICAL

Q2 Callum carried out experiments on water loss from plants. He exposed three plants to different conditions and measured the water loss each time. His results are shown in **Figure 1**.

Figure 1

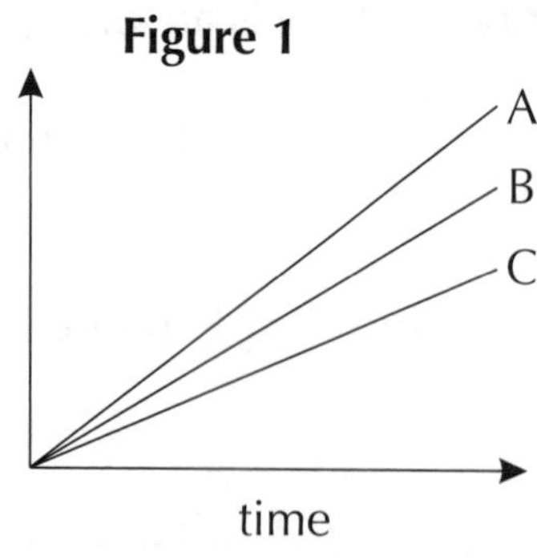

a) Describe the effect that the following conditions will have on the humidity of the air surrounding the leaves.

Remember, air is humid when there's a lot of water in it.

i) Covering the plant with a clear plastic bag.

....

ii) Using a fan to blow air over the plant.

....

b) Complete **Table 1** to show which line (**A**-**C**) from **Figure 1** represents each set of conditions.

Table 1

Conditions	Line
not covered, not fanned	B
covered with clear plastic bag, not fanned	
not covered, fanned	

Q3 Anita is testing the rate of transpiration in a cactus. Cacti live in the desert, where it is very hot during the day but cool at night. The setup of Anita's experiment is shown in **Figure 2**. Her results are shown in **Table 2**. The starting position of the bubble is 0 cm.

Figure 2

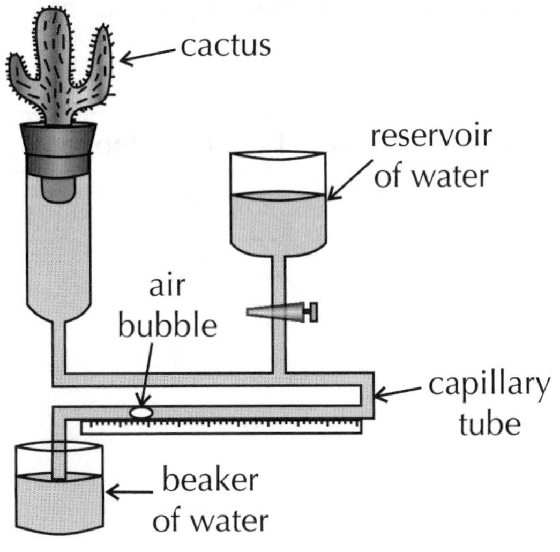

Table 2

Time	Location of bubble (cm)
2 pm	0.1
4 pm	0.2
6 pm	0.2
8 pm	0.3
10 pm	0.7
12 am	1.4
2 am	2.7
4 am	3.6
6 am	4.2
8 am	4.4
10 am	4.5

a) How far does the bubble move during each of the following time periods?

i) 2 pm and 6 pm .... cm

ii) 12 am and 4 am .... cm

iii) 6 am and 10 am .... cm

b) i) What do Anita's results suggest about when most transpiration occurs in a cactus? Tick **one** box.

Use your answers to part a) to help you with b) i).

☐ **A** The amount of transpiration does not change between day and night.

☐ **B** Most transpiration happens during the day.

☐ **C** Most transpiration happens at night.

ii) Suggest how your answer to part **b) i)** affects the amount of water lost by a cactus during transpiration.

Use the information you've been given about deserts in the question intro to help you with b) ii).

....

....

....

Where do you put soldiers in prison? The guard cell...

Stomata comes from the Greek word "stoma", which means mouth. So the underside of leaves are covered in little mouths. That's a creepy image to have in your head — sorry about that. Don't have nightmares about leaves.

Communicable Disease

If we know how diseases are spread, we can help to stop them spreading.

Warm-Up

Pathogens are microorganisms that cause disease. There are different types of pathogen.

Write each of the words in the following list under the correct heading in the table. One has been done for you.

~~rhino~~
bacterium
virus
earthworm
mosquito
fungus
protist
protest

Type of Pathogen	NOT a Type of Pathogen
	rhino

Achoo

Bless you.

Pathogens can be spread in different ways.
Write down three ways in which pathogens can be spread.

1. ....

2. ....

3. ....

Q1 Communicable diseases are caused by pathogens. They can spread from person to person. Non-communicable diseases are not caused by pathogens. They cannot spread from person to person.

Look at the list of diseases below.
Some of the diseases are communicable. Some are non-communicable.

the common cold ....

cancer caused by smoking ....

Type 2 diabetes caused by obesity ....

food poisoning caused by bacteria ....

thrush (an infection caused by a fungus) ....

a) Write '**C**' next to all the diseases in the list that are communicable.

b) Write '**N**' next to all the diseases in the list that are non-communicable.

Q2 Athlete's foot is a skin infection caused by a fungus. It is spread through touching infected skin or surfaces that infected skin has been in contact with (e.g. the floor, towels).

a) Suggest why people with athlete's foot are advised not to walk around with bare feet.

....

....

b) Suggest **one** other step that a person with athlete's foot could take to stop the spread of the disease.

....

Q3 Aisha's class are looking at historical records of a disease outbreak in their town. They are looking at the map in **Figure 1**, which shows where the people who got sick lived. People who were infected had severe diarrhoea and vomiting.

Figure 1

River

House with no infection

House with infection

First case in disease outbreak

Direction of river flow

Aisha's teacher says the disease was probably spread by dirty river water. She says that people got sick from drinking the water.

a) What evidence is there in **Figure 1** that the disease was spread by river water?

....

....

b) Aisha writes down the following suggestions about how the spread of the disease could have been reduced. One of them is incorrect.
Tick the box next to the **incorrect** suggestion.

- [ ] **A** Stop people drinking the river water.
- [ ] **B** Clean up the river water.
- [ ] **C** Destroy vectors of the disease.
- [ ] **D** Vaccinate people against the pathogen that causes the disease.

Bacteria are very communicable — they just never stop talking...

Not all bacteria are pathogens — many are very handy to have around. All viruses are pathogens though.
Now, I don't know about you, but all this talk of pathogens makes me want to go and wash my hands several times.

Types of Communicable Disease

These two pages cover some specific communicable diseases. You need to know how each disease is spread and what symptoms it causes. Use these questions to check you've understood.

Warm-Up

Communicable diseases can be caused by several different types of pathogen. They can be spread in several different ways.

Circle the correct word in each pair to complete the sentences about malaria.

Malaria is a disease caused by a microorganism called a **virus / protist**.

The microorganism is carried between people by **moths / mosquitoes**. An organism which transfers a disease without actually getting it is called a **vector / insect**.

Which of the following are sexually transmitted diseases? Circle the two correct answers below.

Gonorrhoea | Measles | Rose Black Spot | TMV | HIV

Name a disease with all of the following symptoms: red skin rash | fever | cough

....

Q1 Jay ate some chicken that was left out of the fridge overnight. He is now vomiting and has a fever. Jay's doctor thinks he has food poisoning caused by bacteria.

a) Which of these pathogens is likely to be causing Jay's symptoms? Tick **one** box.

☐ **A** Gonorrhoea

☐ **B** *Salmonella*

☐ **C** TMV

b) How is this pathogen causing Jay's symptoms? Tick **one** box.

☐ **A** By producing antigens.

☐ **B** By producing a virus.

☐ **C** By producing toxins.

c) Vaccinating chickens is one way to reduce the spread of the pathogen that has given Jay food poisoning.

Suggest **one** thing that Jay could do at home to reduce the spread of the pathogen.

....

Q2 Tick the boxes to show whether each of the following statements applies to measles or HIV. Some statements apply to both diseases.

		Measles	HIV
a)	This disease is spread by sneezing and coughing.	☐	☐
b)	This disease is caused by a virus.	☐	☐
c)	This disease can eventually cause a patient to develop AIDS.	☐	☐
d)	People with this disease can develop other diseases.	☐	☐
e)	A vaccine for this pathogen has been available for many years.	☐	☐
f)	This disease can be treated with antiretroviral drugs.	☐	☐

Q3 Leah has noticed that some of the plants in her garden look diseased. **Figure 1** shows leaves from two of the infected plants.

Figure 1

A B

a) Draw lines to match the letter of each leaf below to the pathogen it could be infected with.

A	tobacco mosaic virus
B	rose black spot fungus

b) i) What effect do the tobacco mosaic virus and rose black spot fungus have on photosynthesis in the plants they infect?

....

ii) Suggest why both the tobacco mosaic virus and rose black spot fungus reduce plant growth.

....

Q4 Mosquitoes lay their eggs in water. Suggest **one** way that we can use this knowledge to reduce the spread of malaria.

....

....

How do pathogens show their disapproval — they protist...

These diseases are no joke as they can have very serious effects. That's why it's important to know how we can fight diseases and reduce the number of people that get them in the first place. Thankfully we can't catch plant diseases...

Fighting Disease

It's not all bad news if you come into contact with a pathogen — our bodies have some nifty ways to stop them getting inside. And if they do get in, the immune system will fight back.

Warm-Up

The body has lots of features to stop pathogens entering.

How do nose hairs stop pathogens entering the body?
Tick the box next to the correct answer.

- [ ] They produce substances that kill pathogens.
- [ ] They trap particles that could contain pathogens.
- [ ] They produce mucus.

The immune system attacks any pathogens that make it into the body.
White blood cells are part of the immune system.
One way that they attack pathogens is by phagocytosis.

What do white blood cells do during phagocytosis?
Underline the correct answer.

Produce antibodies that stick to a pathogen.

Surround a pathogen and digest it.

Produce toxins that poison the pathogen.

Surround a pathogen and force it to leave the body.

Q1 Different pathogens get into the body in different ways.
The body has different features to try to stop this.

Draw lines to match the pathogens below to the feature of the body that is most likely to stop them entering.

Pathogen	Feature
A bacterium that enters the body through the air and causes lung infections.	cilia and mucus in the bronchi
A bacterium that enters the body in food.	the skin and the substances it produces
A fungus that enters the body through cuts.	stomach acid

Q2 Myra catches mumps, and then she recovers. Later on she is exposed to mumps again. **Figure 1** shows how the level of the antibody against mumps in Myra's blood changes over time.

a) At point **X** on **Figure 1**, Myra was exposed to the mumps virus for the first time. What happens to the antibody level in Myra's blood immediately after point **X**?

....

....

Figure 1

b) Write a '**Y**' on the graph at the point at which Myra was exposed to the mumps virus for the second time.

Q3 John catches chickenpox. For about a week he has an itchy rash all over his skin. He also has a headache and a fever. Then the rash begins to clear up and John starts to feel better.

Don't blame me this time...

a) i) The statements below explain how John's immune system gets rid of the chickenpox virus using antibodies. Write numbers in the boxes to show the correct order of events. The first one has been done for you.

☐ The antibodies allow the chickenpox virus to be found and destroyed by other white blood cells.

1 The chickenpox virus has unique antigens on its surface.

☐ The antibodies lock onto the chickenpox antigens.

☐ When a white blood cell comes across an antigen from a chickenpox virus it starts to make antibodies.

ii) Some white blood cells can produce antitoxins. John's white blood cells **do not** produce antitoxins against the chickenpox virus. Explain why.

....

....

Antitoxins are used to stop toxins from working. Think about why this isn't important for a viral infection.

b) Five years later John is exposed to the chickenpox virus again. Explain why he does not become ill for a second time.

....

....

....

The hairs in my nose are very boring — I wish they were cilia...

White blood cells really are fantastic — we'd struggle to deal with infections without them. Personally, I think phagocytosis is the best way of dealing with enemies — if something gets in your way, just eat it and digest it. Neat.

Fighting Disease — Vaccination

Hopefully these pages don't make you feel all funny if you have a fear of needles. They might be a bit painful to get, but vaccinations have done a lot to reduce the spread of diseases.

Warm-Up

We can help to prevent disease using vaccination. This where a dead or inactive version of a pathogen is injected into the body, causing an immune response.

Tick the correct boxes to show whether the statements about vaccines are true or false.

	True	False
The injected microorganisms have the same antigens as the live pathogen.	☐	☐
White blood cells produce antibodies against the antigens on the injected microorganisms.	☐	☐
After a vaccination, the white bloods cells can produce antibodies to fight all kinds of diseases.	☐	☐
Vaccines can't help to prevent big outbreaks of disease.	☐	☐

Q1 Mia gets injected with the rubella vaccine but Alex doesn't. Soon afterwards they are exposed to the rubella virus.

The passage below explains why Alex gets ill but Mia doesn't.
Use words from the box to fill in the blanks and complete the passage.
Each word can be used once or not at all.

specific	inactive	red blood cells
white blood cells	antigens	antibodies

Mia is protected from infection because her .... can make antibodies to the virus a lot quicker than Alex's can.

When Mia was vaccinated, she was given some ....
rubella pathogens. These had .... on the surface.
Mia's white blood cells then learnt to make antibodies that are
.... to these antigens.

Q2 Polio was a widespread disease in the 20th century. It is caused by a virus and can affect the nervous system. **Figure 1** shows the total number of polio cases every 5 years in the UK from 1916 to 2005.

Figure 1

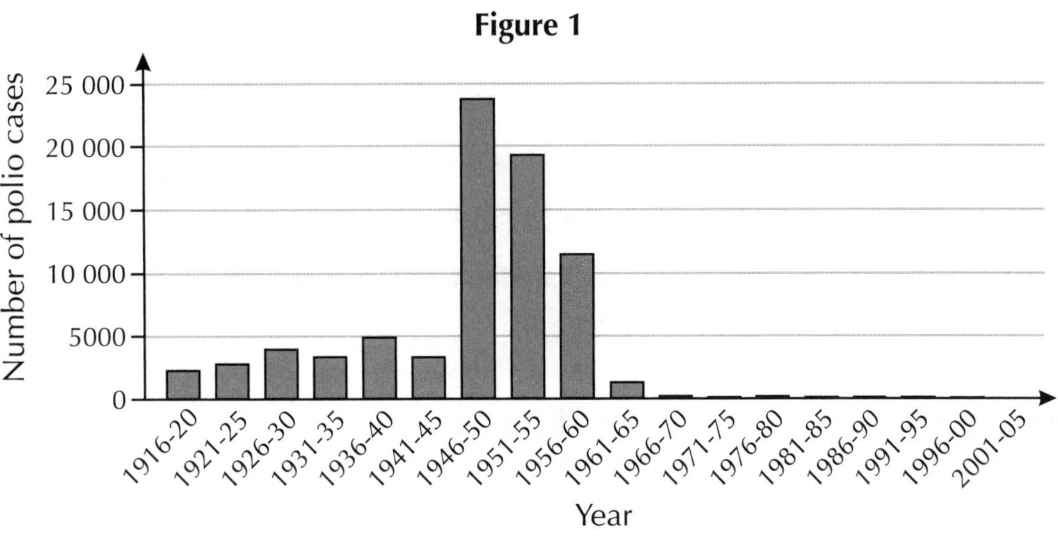

a) Approximately how many cases of polio were there between 1936 and 1940?

....

b) In which 5-year period did the largest outbreak of polio start?
Tick **one** box.

- ☐ **A** 1926 to 1930
- ☐ **B** 1936 to 1940
- ☐ **C** 1946 to 1950
- ☐ **D** 1956 to 1960

c) A vaccine for polio was introduced to the UK in 1956. Do you think the vaccine had an effect on the number of polio cases? Explain your answer.

....

....

d) The last natural infection within the UK was in 1984.
Globally, fewer than 30 cases of polio were recorded in 2017.
Suggest why vaccinations against polio in the UK are still carried out.

....

....

....

....

Hint: use the information you're given in the question.

That's enough about needles for now...

Thanks to vaccination, we've actually got rid of the disease smallpox — the last case anywhere in the world was in the 1970s. Make sure you know how vaccines work to stop people from getting infected and spreading diseases to others.

Drugs and Developing Drugs

Right then, here's another way we can fight diseases when we can't prevent them. Drugs are given to people to cure diseases, or to reduce symptoms.

Warm-Up

Medicinal drugs are used to treat illness. Some drugs attack the pathogen that causes the illness. Other drugs only reduce the symptoms of the illness.

Underline the drug below that reduces symptoms but does not attack pathogens.

antibiotic   painkiller

New drugs often come from plants or microorganisms.
Fill in the missing letters to complete the sentences below.

The painkiller a _ p i r _ _ originally came from w _ l _ o w trees.

The heart medicine d _ _ _ t a l _ _ originally came from _ _ _ g l o _ e _.

The antibiotic _ _ n i c i _ _ _ _ originally came from m _ _ _ d.

Before a new drug can be sold, it has to be tested to make sure it works and is safe. The drug is tested on different cells and organisms.

Write a number from 1 to 4 in each of the boxes to put the cells and organisms in the order that drugs would be tested on. The first one has been done for you.

- [1] Human cells in a laboratory
- [ ] Healthy human beings
- [ ] Patients with the disease
- [ ] Mammals (other than humans)

Q1 An online advert for a new pill states that taking it can reduce 'bad' cholesterol by 52% (compared with 7% using a placebo). The new pill was tested in a drug trial that used a placebo.

a) What could the placebo have been in this drug trial?
Tick **one** box.

- [ ] **A** An injection that reduces 'bad' cholesterol.
- [ ] **B** An injection that doesn't do anything.
- [ ] **C** A pill that looks like the pill that reduces 'bad' cholesterol but doesn't do anything.
- [ ] **D** A different pill that also reduces 'bad' cholesterol.

b) Why did the manufacturer use a placebo when testing the new pill?
Tick **one** box.

- ☐ **A** To make sure the new pill was safe to use.
- ☐ **B** It was too expensive to give everyone the new pill.
- ☐ **C** To find out what the dose of the new pill should be.
- ☐ **D** To find out whether the new pill is really effective.

c) The patients involved in the drug trial were not told whether they were getting the placebo or the new pill. The doctors involved in the trial did not know either.

What name is given to this type of trial? Circle the correct answer below.

blind trial **double-blind trial** **triple-blind trial**

d) Do you think the new pill is effective? Justify your answer.

'Justify your answer' means support your answer with evidence (e.g. from the information given to you in the question).

....

....

....

Q2 A new medicine called 'Killcold' contains painkillers and decongestants to treat colds. Painkillers help to relieve the pain of a sore throat or headache. Decongestants help you to breathe more easily when you have a blocked nose. Colds are caused by viruses.

KILL COLD

a) Explain why the new medicine's name isn't strictly accurate.

....

....

b) Why don't doctors give antibiotics for colds?

....

....

c) Suggest why it is difficult to develop a drug to destroy the cold virus.

....

....

....

Double-blind marking — when neither you nor your teacher knows the answers...

A lot of drugs in the past were based on substances found in plants, and scientists still look at plants now to find new ones. But if you have a headache, don't ever just chew random plants in the garden — some plants are pretty toxic. That's one of the reasons why drugs are tested so carefully as part of drug trials — to make sure they're safe.

Photosynthesis

Photosynthesis is a scary word. But don't worry, it's just how plants make food...

Warm-Up

Plants use light to make glucose. This is called photosynthesis.

Circle the correct word in each pair to complete the sentences below.

Photosynthesis takes place in the **chloroplasts / nucleus** of plant cells. It uses light energy to change carbon dioxide and **water / oxygen** into glucose and **water / oxygen**.

Photosynthesis is an **endothermic / exothermic** reaction.

Only one of the following statements about photosynthesis is true. Tick the correct one.

- [ ] **A** Photosynthesis happens all day and all night in plant cells.
- [ ] **B** Photosynthesis involves energy transfer from the chloroplasts to the environment.
- [ ] **C** Photosynthesis needs chlorophyll, which is found in the cell nucleus.
- [ ] **D** Photosynthesis involves energy transfer from the environment to the chloroplasts.

Q1 Potatoes store their glucose as starch.
New potato plants can be grown from potatoes.

a) The sentences below explain why potatoes store their glucose as starch. Complete the sentences by circling the correct word from each pair.

Starch is **soluble / insoluble**. This means it **causes / doesn't cause** water to be drawn into potato cells by osmosis.

b) Whilst below ground, new potato plants break down the starch in potatoes into glucose. Suggest **two** ways that the new potato plants use this glucose.

1. ....

2. ....

c) When the new potato plants grow above ground, they don't need to break down starch stores to get glucose. Suggest how the new plants get their glucose instead.

....

I'm feeling quite peckish — I could do with a light snack...

Photosynthesis is a plant's way of knocking up a roast dinner. Well, not quite. But it's a pretty clever process. Make sure you know the main ways that plants use the glucose they produce. Then turn the page for more on photosynthesis.

The Rate of Photosynthesis

Some things can speed up the rate of photosynthesis, other things can slow it down. Such is life.

Warm-Up

Light intensity, temperature, carbon dioxide concentration and the amount of chlorophyll in a plant are all limiting factors for photosynthesis.

Fill in the gaps in the passage below using some of the words in the box.

faster	too little	carbon dioxide	slower	chlorophyll	too much

A limiting factor is something that stops photosynthesis from happening any .... Chlorophyll can be a limiting factor of photosynthesis if a plant has .... chlorophyll. When this happens, the rate of photosynthesis can only increase if there is an increase in ....

Q1 Farmer Fern doesn't put her cows out during the winter because the grass is not growing.

a) Suggest **two** differences between summer and winter conditions that could affect the rate of photosynthesis in the grass.

1. ....

2. ....

b) How are the rate of photosynthesis and the growth rate of grass related? Tick **one** box.

☐ **A** When the rate of photosynthesis increases, the growth rate of grass increases.

☐ **B** When the rate of photosynthesis increases, the growth rate of grass decreases.

☐ **C** When the rate of photosynthesis decreases, the growth rate of grass increases.

Q2 Hannah is growing some herbs in her kitchen.

a) Hannah switches on a light bulb in her kitchen.
What will happen to the rate of photosynthesis after she turns on the light?
Give a reason for your answer.

The rate of photosynthesis will .... **because** ....

....

b) Hannah switches on an electric heater in her kitchen.
Suggest why the rate of photosynthesis in her herbs increases.

....

....

c) Hannah buys more herbs to grow in her kitchen.

i) How do you think increasing the number of herbs will affect the carbon dioxide concentration in Hannah's kitchen? Underline the correct answer below.

The carbon dioxide concentration will **increase / decrease / stay the same.**

ii) How do you think increasing the number of herbs in the kitchen will affect the rate of photosynthesis of the herbs?

Use your answer to c) i) to help you.

....

....

PRACTICAL

Q3 Levi measured the rate of photosynthesis in pondweed at different temperatures. His results are shown in **Figure 1**.

Figure 1

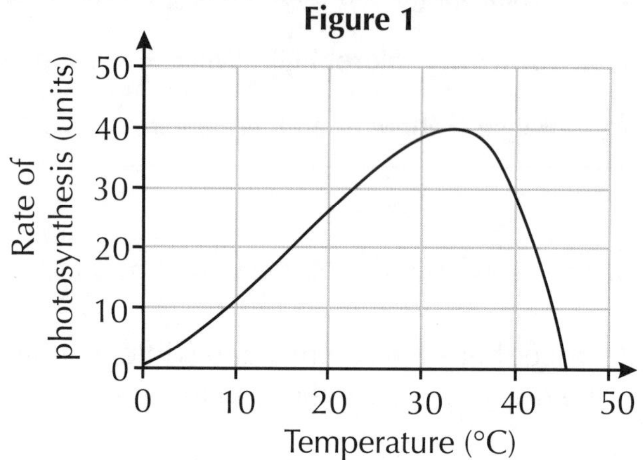

a) i) What is the highest rate of photosynthesis in **Figure 1**?

.... units

ii) Approximately what temperature produces the highest rate of photosynthesis?

.... °C

iii) If the temperature isn't right, the enzymes needed for photosynthesis can be denatured. At what temperature are the enzymes denatured in **Figure 1**?

.... °C

b) i) Suggest **one** variable Levi might have measured in order to calculate the rate of photosynthesis in the pondweed.

....

ii) Describe **one** way that Levi could have measured the variable you suggested in part **b) i)**.

....

....

Time spent on the internet slows my work rate right down...

So the temperature has to be just right — photosynthesis may stop if it's too hot or too cold. There also has to be enough light, enough carbon dioxide and enough chlorophyll. Phew, I wonder if the plants have any more demands...

Respiration and Metabolism

Respiration and metabolism — two fancy words for two very important processes...

Warm-Up

Respiration is the process of transferring energy from glucose.
Respiration can be aerobic or anaerobic.

Use the words given to complete the word equation for aerobic respiration.

I told you aerobic respiration transferred lots of energy.

Do you think anyone will notice I blew up the house?

oxygen   water   glucose

.... + ... → carbon dioxide + ... (+ energy)

Which of the following is the chemical symbol for glucose? Circle the correct answer.

CO_2 $C_6H_{12}O_6$ O_2 H_2O

Which statements about respiration are true? Tick two boxes.

- [ ] **A** Respiration only happens in some of your body cells, some of the time.
- [ ] **B** Respiration is an endothermic reaction that takes in energy.
- [ ] **C** Respiration is an exothermic reaction that transfers energy to the environment.
- [ ] **D** Respiration provides energy for all living processes.

Q1 Metabolic reactions are chemical reactions that occur in all living organisms.

a) Which metabolic reaction listed below happens in plants, but **not** in humans? Tick **one** box.

- [ ] **A** Glucose is broken down in respiration.
- [ ] **B** Lipid molecules are made from glycerol and fatty acids.
- [ ] **C** Glucose molecules are joined together to form cellulose.

b) Metabolic reactions could stop happening if an organism gets too hot. Suggest why.

Hint: think about how metabolic reactions are controlled.

....

....

Q2 Mice use both aerobic respiration and anaerobic respiration.

a) i) How does the energy transferred by aerobic respiration help mice when they are in cold environments?

....

ii) Write down **two** other things that aerobic respiration helps mice to do.

1. ....

2. ....

b) i) When mice respire anaerobically, do they transfer more or less energy than when they respire aerobically?

....

ii) Suggest why mice are unable to survive using anaerobic respiration alone.

....

iii) Give the word equation for anaerobic respiration in a mouse's muscle cells.

....

PRACTICAL

Q3 Priyanka sets up the equipment in **Figure 1** to investigate respiration in germinating seeds. Germinating seeds produce a gas when they respire. The gas bubbles into the limewater and makes it cloudy.

Figure 1

test tube containing air

germinating seeds

limewater

a) i) Identify the gas produced by the germinating seeds.

....

ii) Give another product made by the germinating seeds.

....

b) Priyanka notices that the test tube containing the seeds feels warm to touch. Suggest why.

....

....

c) Suggest how Priyanka could show that the gas was produced by the seeds respiring and not simply the presence of seeds in the test tube.

....

Be thankful — respiration gave you the energy to do these questions...

Remember, respiration isn't breathing in and out. Respiration is how the energy you get from food is transferred so it can be used for all kinds of things. Handy. Respiration can be aerobic or anaerobic — make sure you know the difference.

Exercise

On your marks, get set, start this page of questions on exercise...

Warm-Up

When you exercise you <u>respire more</u>. This transfers <u>more energy</u> to your muscle cells. Your body also responds to get <u>more oxygen</u> to your muscle cells.

Write down <u>one</u> way your body responds to get more oxygen to your muscle cells.

....

Q1 Jim runs a race. **Figure 1** shows Jim's breathing rate before, during and after the race.

Figure 1

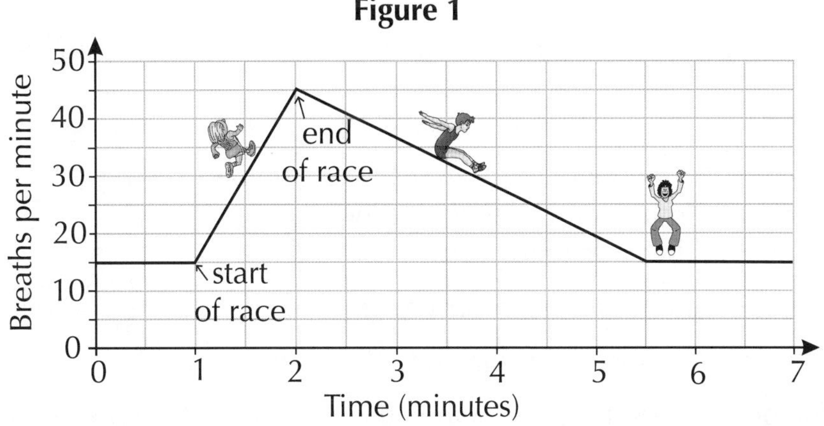

a) At the start of the race Jim's breathing rate is 15 breaths per minute.
How many breaths per minute does Jim's breathing rate go up by during the race?

.... breaths per minute

b) How long does it take for Jim's breathing rate to return to normal after the race?
Tick **one** box.

☐ **A** 5.5 minutes ☐ **B** 4.5 minutes ☐ **C** 3.5 minutes

c) How would you expect Jim's breath volume to change during the race?
Explain your answer.

....

....

d) Near the end of the race, Jim's muscles start to get tired. Suggest why.

....

....

Phew, this topic gave my brain a work-out...

Remember that exercise makes you respire more. If you do really hard exercise, you start to respire anaerobically and build up an oxygen debt. Now that you've finished this page, give yourself some time to recover before starting Topic B5.

Homeostasis

Homeostasis is how your body keeps things ticking over smoothly.

Warm-Up

<u>Homeostasis</u> means keeping the conditions inside your body and cells nice and <u>steady</u>. It involves <u>automatic control systems</u>. Which of these statements about automatic control systems is <u>true</u>?

- [ ] **A** Automatic control systems involve hormones only.
- [ ] **B** Automatic control systems involve the nervous system only.
- [✓] **C** Automatic control systems can involve either hormones or the nervous system.

Q1 Blood pressure is an example of a condition in the body that needs to be kept steady.

The statements below describe what happens when blood pressure gets too high.

a) Write each of the numbers **1** to **4** in the boxes below to show the order the statements should be in. The first one has been done for you.

If you don't know anything about blood pressure, it doesn't mean you can't answer these questions. Read the information you're given and think about what you know about how control systems work.

[3] The brain processes the information.

[2] The receptors send information to the brain.

[4] The brain sends impulses to the heart, which starts to beat slower. This helps to return blood pressure to normal.

[1] An increase in blood pressure is detected by receptors in the blood vessels.

b) i) What is the stimulus that causes the response described above?

Getting a Jump ✗

ii) What is the effector in the response described above?

Heart ✓

c) The brain is the coordination centre in this response. What does this mean?

It sends impulses to the right part of the body. ✓

d) If blood pressure falls too low, what will this control system do? Tick **one** box.

- [✓] **A** Automatically cause blood pressure to increase.
- [ ] **B** Automatically cause blood pressure to decrease.
- [ ] **C** Nothing.

Good old homeostasis — no one likes change...

Homeostasis means that even if conditions outside of your body change, everything in your body continues to work as it should. This happens automatically, without you even having to think about it. Shame learning isn't like that.

The Nervous System, Synapses and Reflexes

These intros are really starting to get on my nerves. I'm never quite sure how to react to them...

Warm-Up

The nervous system detects and reacts to stimuli (changes in the environment). Information passes along neurones to the central nervous system (CNS), which coordinates the response.

Draw arrows between the boxes in the diagram to show the flow of information from a stimulus through the nervous system to the response. The first two have been done for you.

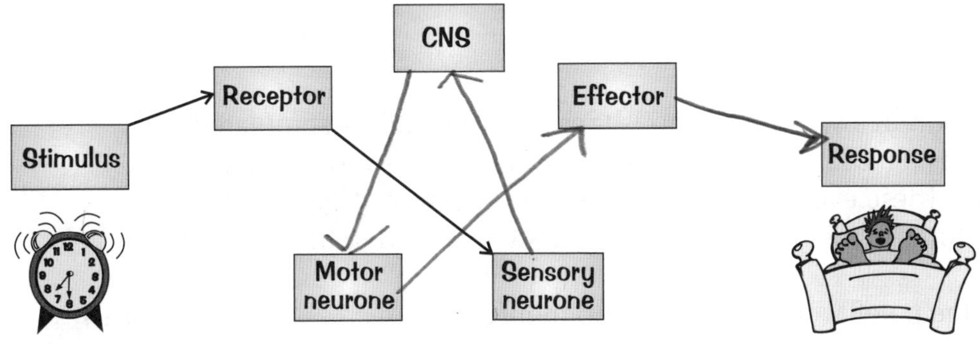

Which parts of the body make up the CNS?

Brain spinal cord

Q1 Reflex actions are automatic responses to a stimulus.

The two situations below would cause reflex actions.

A: Stepping on a drawing pin with bare feet.
B: Smelling food when hungry.

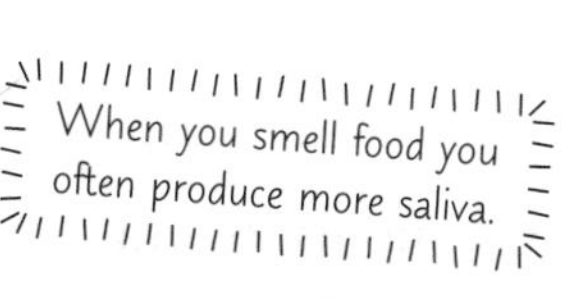

Use the words in the box to complete **Table 1** for situations **A** and **B**.

muscles in the leg | pressure receptors | smell of the food | smell receptors | saliva released into mouth

Table 1

	A	B
stimulus	pressure of the pin	Smell of the food
receptor	pressure receptor	Smell receptors
effector	muscles in the leg	salivary glands
response	foot is lifted up	Saliva released into mouth.

Q2 Jamie has just taken a delicious-looking lasagne out of the oven. Unfortunately he forgot the dish would still be hot and picked it up without oven gloves on. This caused him to drop the dish on the floor. Jamie dropping the lasagne is an example of a reflex reaction.

a) Tick the box next to **all** the sentences below that you think are **true**.

☐ Jamie dropped the lasagne immediately.

☐ Jamie dropped the lasagne without thinking about it.

☐ The neurones involved in dropping the lasagne went through a conscious part of Jamie's brain.

☐ It took Jamie a few seconds to drop the lasagne.

☐ Jamie thought about whether or not he should drop the lasagne.

b) What is the purpose of reflex reactions, such as dropping hot lasagnes?

....

Q3 If the tendon below the kneecap is tapped, a receptor is stimulated. This causes a muscle in the thigh to contract and the lower leg to kick upwards. These events are part of the patellar reflex. **Figure 1** shows the structures in the nervous system that are involved in this reflex arc.

a) In the patellar reflex, what sort of neurone carries the information from:

i) the receptor to the spinal cord?

Sensory neurone

ii) the spinal cord to the thigh muscle?

Motor nerone

b) What type of neurone is missing from this reflex arc?

relay neurone

Figure 1

receptor

X

muscle in thigh

spinal cord

c) Complete the following sentence.

In this reflex arc, the muscle causing the leg to kick acts as the ....

d) i) What is the name of the junction marked **X** on **Figure 1**?

Synapse

ii) Explain how the nerve signal is transferred across junction **X**.

passes an electrical / Chemical Sygnial to another nerone

With great reflexes comes great response-ability...

We need to be able to respond to changes in our external environment in order to survive, and our nervous system allows us to do just that. We can react to our surroundings and coordinate our behaviour accordingly. Clever stuff.

The Endocrine System

The endocrine system is just a fancy way of describing the hormonal system.

Warm-Up

Hormones are chemical messengers sent in the blood. They are produced in endocrine glands. Each hormone only affects particular cells in target organs.

Use the words below to complete the diagram showing human endocrine glands.

ovary

pituitary gland

pancreas

adrenal gland

thyroid

pituary gland ....

..........Thyroid..........

..........adrenal gland..........

Ovary ....

..........pancreas..........

Q1 Tick the boxes to show whether the following responses are mainly controlled by the nervous or hormonal systems.

		Nervous system	Hormonal system
a)	Hearing the alarm clock and turning it off.	✓	
b)	A child growing taller.		✓
c)	Smelling toast burning.	✓	
d)	The development of sexual characteristics during puberty.		✓

Q2 One of the endocrine glands in the body is sometimes called the 'master gland'. It's called this because it produces hormones that act on other endocrine glands.

Some diseases can stop the 'master gland' from working normally. Suggest why these diseases may have many different effects on the body.

....

....

....

Your brain — the target organ for these questions...

The endocrine system is different to the nervous system, but it provides another way to send information round the body. Hormones tend to have quite long-lasting effects and control things in organs that need constant adjustment.

Controlling Blood Glucose

Your blood glucose level needs to be kept in check to prevent serious problems.

Warm-Up

Insulin is a hormone. It controls the blood glucose level.

Fill in the gaps in the passage below using words from the box.
You don't need to use all of the words. You may use one word more than once.

pancreas   liver   glycogen   insulin

The .... monitors blood glucose concentration. Insulin is produced in the .... Glucose can be stored as ....

Diabetes is a condition where you can't control your blood glucose level.
There are two main types of diabetes — Type 1 and Type 2.

Which statements about Type 1 diabetes are true? Tick two boxes.

- ☐ **A** Insulin is produced, but body cells do not respond properly to it.
- ☐ **B** Little or no insulin is produced.
- ☐ **C** Insulin injections help to increase the amount of glucose in the blood.
- ☐ **D** Insulin injections help to reduce the amount of glucose in the blood.

Q1 Ruby has Type 1 diabetes.

a) Ruby eats a meal containing carbohydrate. This makes her blood glucose level rise.

i) Why does eating the meal cause Ruby's blood glucose level to rise?

....

ii) Ruby injects insulin before every meal. The insulin stops Ruby's blood glucose level from getting too high. How does it do this?

....

....

b) i) Ruby starts playing football three times a week.
How is playing a game of football likely to affect her blood glucose level?
Tick **one** box.

- ☐ **A** Her blood glucose level is likely to rise.
- ☐ **B** Her blood glucose level is likely to fall.
- ☐ **C** Her blood glucose level is likely to stay the same.

ii) Do you think Ruby is likely to need more or less insulin on the days she plays football? ....

Q2 **Table 1** shows some statistics about diabetes in the UK that were reported in 2016.

Table 1

Number of people with diabetes under the age of 19		
Boys	Girls	Total
16 380	15 120	31 500

a) i) What percentage of people with diabetes under the age of 19 were boys?

.... %

ii) It is estimated that around 95% of the total shown in **Table 1** have Type 1 diabetes. How many people in **Table 1** are likely to have Type 1 diabetes?

.... people

b) Most adults who are diagnosed with diabetes have Type 2 diabetes. Suggest **one** reason why Type 2 diabetes might be more common in adults than in children.

....

....

Q3 Some pregnant women develop a condition called gestational diabetes. They produce insulin but their body cells stop responding to it.

a) Is gestational diabetes more similar to Type 1 or Type 2 diabetes? Give a reason for your answer.

Type .... **diabetes because** ....

....

....

Don't panic if you haven't heard of gestational diabetes. Compare the information you've been given with what you know already about diabetes.

b) Gestational diabetes can be controlled by eating a carbohydrate-controlled diet. Suggest **one** other way gestational diabetes could be controlled.

....

I'll keep this summary short and sweet...

The hormone insulin is very important for keeping your blood glucose level steady. People with diabetes have problems producing or responding to insulin. It is a serious condition that has to be carefully controlled to prevent complications.

Puberty and the Menstrual Cycle

A lot of changes happen in the body at puberty, including the start of the menstrual cycle in women.

Warm-Up

An egg is released from a woman's ovaries every month as part of the menstrual cycle.

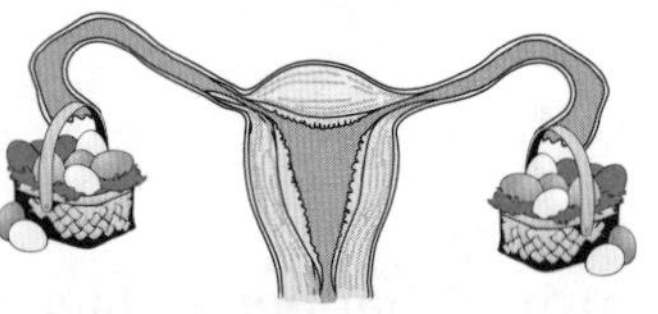

The menstrual cycle is controlled by four hormones.
Place ticks in the correct boxes below to show the role of each of these hormones.

Role	FSH	LH	Progesterone	Oestrogen
Causes an egg to mature in an ovary.	✓			
Causes the release of an egg.		✓		
Involved in the growth and maintenance of the uterus lining.			✓	✓

Q1 David is 11. He has just started going through puberty.

a) Name the reproductive hormone that will cause David's voice to get deeper.

testerone

b) Which glands in David's body produce this hormone?

testes

c) Suggest **one** other effect of this hormone on David's body.

facial hair

Q2 Different hormones control different parts of the menstrual cycle.
If the level of a hormone is too low, a woman's menstrual cycle can be affected.

Draw lines to match each problem below with the hormone that might be responsible.

Problem	Hormone responsible
An egg matures in the ovaries but ovulation does not take place.	LH
An egg does not mature in the ovaries.	progesterone
Bleeding occurs between periods.	FSH

Q3 Joanna's menstrual cycle lasts 35 days. How much longer is this than the average menstrual cycle? Give your answer as a percentage.

25 %

Q4 Very rarely, a young child may have a medical condition that means her ovaries need to be removed. In these cases, the child will be prescribed a hormone by her doctor, so that she is still able to go through puberty. Suggest which hormone the doctor is likely to prescribe.

Oestrogen

Q5 The menstrual cycle has four main stages.

a) The statements below describe what happens during the menstrual cycle. Write the numbers **1** to **4** in the boxes below to put the statements in the correct order. The first one has been done for you.

[2] A thick spongy layer of blood vessels forms as the uterus lining builds up.

[4] The uterus wall stays the same for about two weeks.

[**1**] The uterus lining breaks down for about four days, causing bleeding.

[3] An egg is released after maturing in the ovary.

b) i) Suggest which of the statements from part **a)** describes the stage of the menstrual cycle where the level of LH will be highest.

3

ii) Explain your answer to part **b) i)**.

Because LH is responsible for releasing the egg

c) The progesterone level decreases just before the stage of the cycle marked '**1**' in part **a)**. Explain why the progesterone level needs to decrease at this point.

because progesterone is responsible for thickening the uterus

There's nothing funny about the menstrual cycle ***period***

People used to think that the moon controlled a woman's menstrual cycle. (Yes, you read that right). We now know that the menstrual cycle is controlled by hormones, which is pretty cool. Not as exciting as the moon thing though.

Controlling Fertility

Different forms of contraception can be used to stop a woman getting pregnant.

Warm-Up

<u>Contraceptives</u> are used to <u>prevent pregnancy</u>.

Tick the correct boxes to show whether the contraceptives below are <u>hormonal</u> or <u>non-hormonal</u>.

	Hormonal	Non-hormonal
Contraceptive injection	✓	
Diaphragm		✓
Abstinence		✓
Contraceptive patch	✓	
Condom		✓

Q1 Intrauterine devices (IUDs) are contraceptive devices that are inserted into the uterus.

a) Like contraceptive implants, plastic IUDs release the hormone progesterone. Suggest **one** way that the release of progesterone by a plastic IUD prevents pregnancy.

Stop releasing egg

b) Copper IUDs kill sperm. Give **one** other form of contraception with the same function.

Spermicide

Q2 Different forms of contraception are suitable for different people.

a) Candice wants a hormonal contraceptive that she doesn't have to think about every day or every time she has sex. Suggest **one** type of contraception that she could use.

Sterilisation

b) Liza and Amir would like to have children one day, but not at the moment. Give **one** reason why sterilisation is **not** a suitable choice of contraception for them.

Because it's very hard to undo

This stuff is to be taken seriously you know...

Hormones can be used in contraceptives, but there are plenty of contraceptive methods that don't use hormones. Take the time to learn the different methods and make sure you know the pros and cons of each type.

DNA

DNA carries all of your genetic information. If you're thinking it's probably quite important, then you'd be right. Answer these questions like your life depends on it...

Warm-Up

The genome is all the genetic material in an organism.
DNA is the chemical that all genetic material is made from.

Tick the correct box to complete each of the following sentences.

DNA is found in really long structures called...

...chromosomes ☐ ...ribosomes ☐

A DNA molecule is made up of strands of DNA coiled together, making a...

...single helix ☐ ...double helix ☐ ...triple helix ☐

Sections of DNA that code for a particular sequence of amino acids are called...

...proteins ☐ ...genomes ☐ ...genes ☐

Q1 Write out the structures below in order of size, starting with the smallest.

nucleus gene chromosome cell

1. .... 2. .... 3. .... 4. ....

Q2 Scientists have identified a gene that is linked to breast cancer. Individuals who inherit an altered form of this gene have more chance of developing breast cancer.

a) Suggest how scientists have identified genes that are linked to specific diseases, such as breast cancer.

....

....

b) Suggest **one** advantage of knowing that a specific gene is linked to breast cancer.

....

Just follow the DNA sequence — see where life takes you...

Here's a fun fact for you — over half of our DNA is shared with bananas. I'm not joking. Scientists found this out by studying the genetic material of organisms. (Don't worry. They also learnt many more useful things besides this.)

Reproduction and Meiosis

Sexual reproduction usually involves two organisms. It doesn't always take two to ~~tango~~ reproduce though — only one parent is required for asexual reproduction.

Warm-Up

Asexual reproduction produces genetically identical cells.
Sexual reproduction produces genetically different cells.

Complete the passage below about sexual reproduction by circling the most appropriate word or words in each pair.

Sexual reproduction involves the **fusion / replication** of gametes. The gametes are made by **mitosis / meiosis**. They have **half the / the same** number of chromosomes as normal cells.

Q1 Cells divide by meiosis to form gametes. Draw lines to match the descriptions of this process to the correct diagrams below. The first one has been done for you.

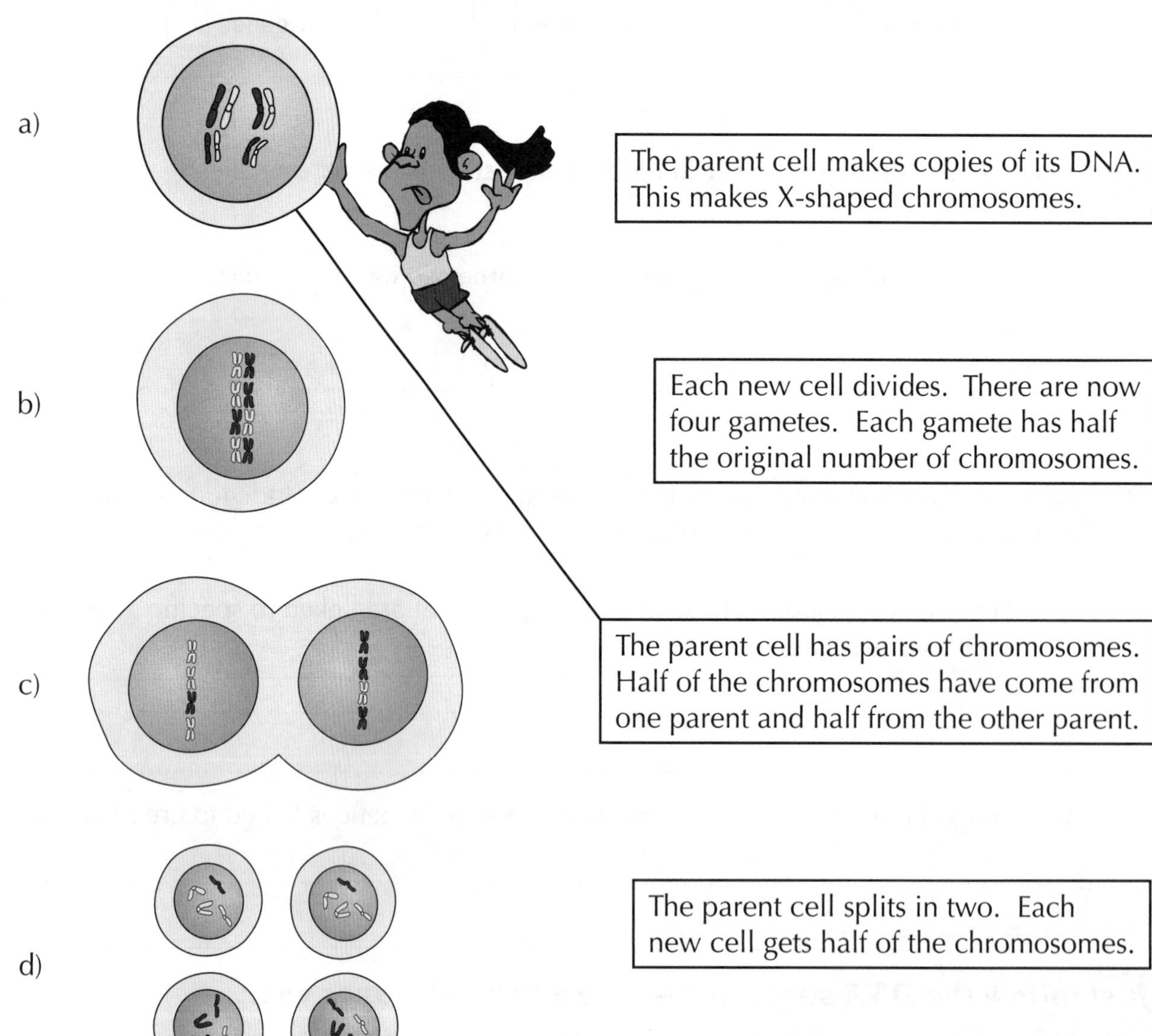

Q2 For each of the statements below, tick **one** box to show whether what they describe relates to sexual reproduction or asexual reproduction.

		Sexual reproduction	Asexual reproduction
a)	A daughter yeast cell is genetically identical to the parent yeast cell.	☐	☐
b)	Female salmon lay eggs and male salmon cover them in sperm.	☐	☐
c)	Pollen and egg cells join together to form new sunflowers.	☐	☐
d)	In some species of lizard, females can reproduce without males.	☐	☐
e)	Dogs produce gametes by meiosis.	☐	☐
f)	The offspring of a cat all look different.	☐	☐
g)	Bacteria reproduce by splitting in two.	☐	☐
h)	If a flatworm is cut into pieces, each piece can grow into a new flatworm using mitosis.	☐	☐

Q3 Mice reproduce using sexual reproduction.

a) Mouse body cells have 40 chromosomes. The chromosomes are found in pairs. Mice produce gametes by meiosis.

i) How many chromosomes do mouse gametes have? Tick **one** box.

☐ **A** 40
☐ **B** 80
☐ **C** 20
☐ **D** 10

ii) Mouse egg and sperm cells join together at fertilisation. How many chromosomes are in the fertilised egg?

.... chromosomes

b) i) A fertilised mouse egg divides many times to form a mouse embryo. What kind of cell division takes place to form a mouse embryo?

....

ii) How do the cells in the embryo change as the mouse develops?

....

Mitosis, meiosis, Brussels sprouts — very divisive issues...

Mitosis and meiosis sound really similar — make sure you don't get them confused. Mitosis produces cells that are genetically identical to the parent cells. Meiosis produces cells that are genetically different to the parent cells. Simple...

 ☐  ☐  ☐

X and Y Chromosomes

Here's a page of questions on sex chromosomes — that must be whY I'm so eXcited...

Warm-Up

A person's sex is determined by a pair of chromosomes — X and Y.

Q1 **Figure 1** shows the chromosomes of two people.

Figure 1

Person A

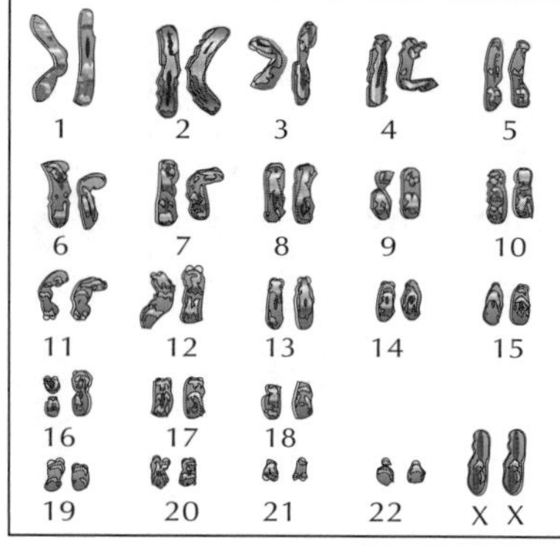

Person B

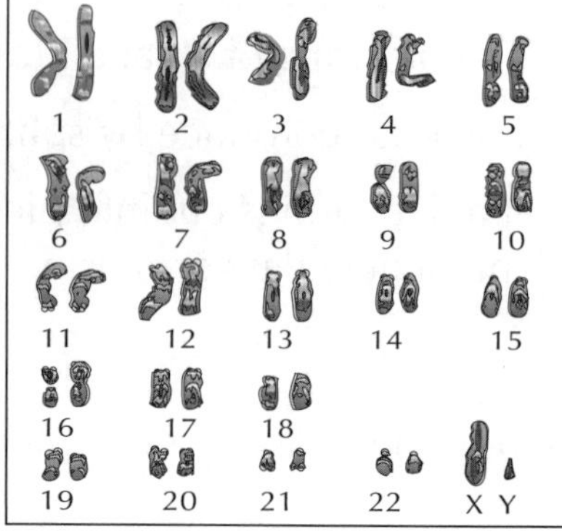

Which person is male? Give a reason for your answer.

Person .... is male because ....

Q2 Birds have sex chromosomes called **Z** and **W** (just like humans have X and Y). Birds with two **Z** chromosomes (**ZZ**) are **male**. Birds with the chromosomes **ZW** are **female**.

a) Complete the Punnett square shown in **Figure 2**.

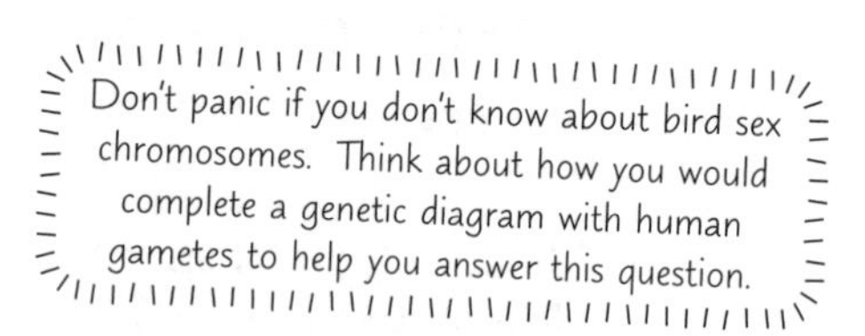

Don't panic if you don't know about bird sex chromosomes. Think about how you would complete a genetic diagram with human gametes to help you answer this question.

Figure 2

male gametes \ female gametes	Z	W
Z	ZZ	
Z		

b) What is the probability of getting a male bird in the cross in **Figure 2**?

....

I wonder what all the other chromosomes are called...

You just never know when you might need to carry out a genetic cross — luckily for you there's more practice coming up.

Genetic Diagrams

Genetic diagrams can be used to show how characteristics are inherited.

Warm-Up

Different genes control different characteristics.
Most characteristics are controlled by several genes.

Some characteristics are controlled by a single gene. Write down one example.

....

Alleles are different versions of a gene. You have two alleles for each gene in your body. One comes from your mum. The other comes from your dad.

Draw lines to match the type of allele to the correct description and then to the type of letter used to represent it.

Dominant allele	The allele whose characteristic is only shown if two copies are present.	small letter, e.g. 'a'
Recessive allele	The allele whose characteristic is shown even if only one copy is present.	capital letter, e.g. 'A'

Q1 In rabbits, the recessive allele (**f**) produces solid fur colour.
The dominant allele (**F**) gives agouti fur colour.
Agouti fur colour has a banded pattern.

a) i) A rabbit has the alleles **Ff**.
Does this rabbit have solid fur colour or agouti fur colour?
Give a reason for your answer.

The rabbit has .... **fur colour because** ....

....

ii) Is the rabbit homozygous or heterozygous for this characteristic?

....

b) i) What are the possible genotypes for rabbits with agouti fur colour?
Tick **two** boxes.

☐ **A** ff ☐ **B** F ☐ **C** Ff ☐ **D** FF

ii) What genotype would a rabbit with solid fur colour have?
Tick **one** box.

☐ **A** ff ☐ **B** f ☐ **C** Ff ☐ **D** FF

Q2 Seeds of pea plants can be yellow or green. The allele for yellow seeds (**Q**) is dominant. The allele for green seeds (**q**) is recessive.

a) What are the possible genotypes of pea plants with yellow seeds?
Circle **two** correct answers.

QQ **qq** **Qq**

b) i) **Figure 1** shows a cross between a pea plant with genotype **QQ** and a pea plant with genotype **qq**. Complete **Figure 1** to show the offspring's genotypes.

Figure 1

ii) What proportion of the offspring in **Figure 1** will have yellow seeds?
Tick **one** box.

☐ **A** 100% ☐ **B** 50% ☐ **C** 25% ☐ **D** 0%

c) i) Two pea plants (**Qq**) are crossed.
Complete the Punnett square in **Figure 2** to show this cross.

Figure 2

		parent's alleles	
		Q	q
parent's alleles	Q		
	q		

ii) Circle any offspring in **Figure 2** that will have green seeds.

Move over crosswords — completing genetic diagrams is my new hobby...

Genetic diagrams look really confusing at first. But take the time to understand how they work and you'll be knocking them out in no time at all. Remember that they only show probability — not what will definitely happen.

Inherited Disorders and Family Trees

Some more genetic diagrams coming up now — practice makes perfect after all.

Warm-Up

Some genetic disorders can be inherited from your parents.

Embryos can be screened for inherited disorders.

Write down one reason why some people think embryo screening is a bad thing.

....

....

Q1 Polydactyly is a genetic disorder. It is caused by a dominant allele.
The dominant allele for polydactyly is **D**, and the recessive allele is **d**.

a) A baby is born with two thumbs and nine fingers.
Could this baby have polydactyly? Give a reason for your answer.

....

b) A person has the alleles **Dd**.
Will this person have polydactyly, be a carrier or be unaffected?
Explain your answer.

The person will .... because ....

....

....

c) i) Complete the genetic diagram in **Figure 1** showing the inheritance pattern of polydactyly.

Figure 1

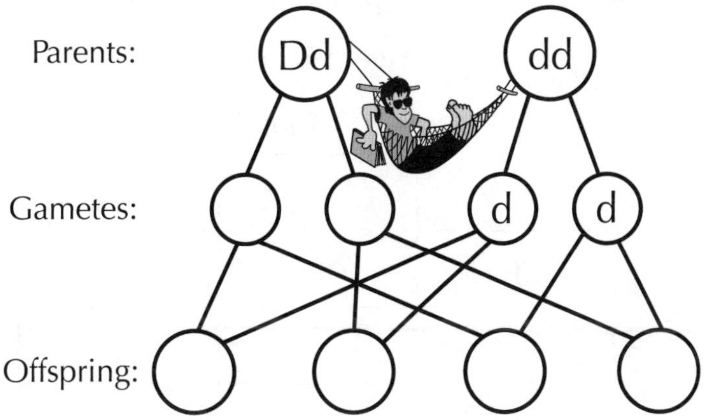

Parents: Dd dd

Gametes: d d

Offspring:

ii) In **Figure 1**, what proportion of the offspring will have polydactyly?
Tick **one** box.

☐ **A** 1 ☐ **B** 0.5 ☐ **C** 0.25 ☐ **D** 0

Q2 The family tree in **Figure 2** shows a family with a history of a genetic disorder.

Figure 2

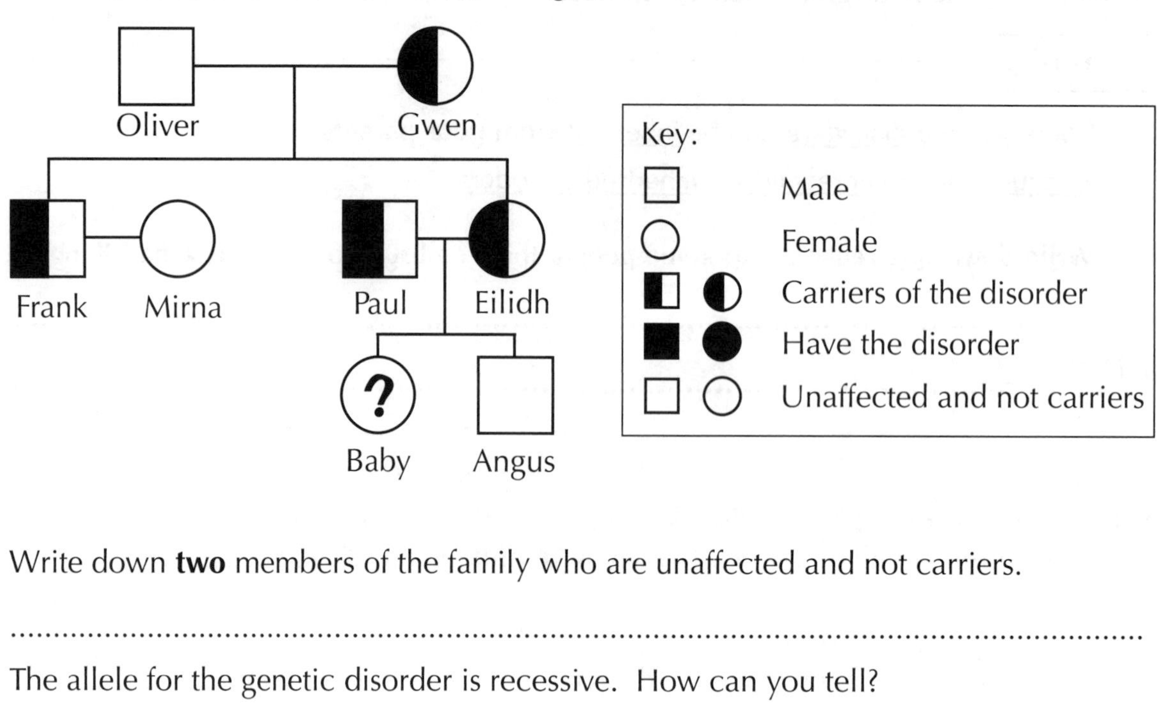

a) Write down **two** members of the family who are unaffected and not carriers.

....

b) The allele for the genetic disorder is recessive. How can you tell?

....

....

c) i) The recessive allele for this disorder is **f**. The dominant allele is **F**.
Paul is a carrier of the disorder. What is Paul's genotype?
Tick **one** box.

☐ **A** FF

☐ **B** Ff

☐ **C** ff

☐ **D** f

ii) Paul and Eilidh are having a baby. Complete the Punnett Square in **Figure 3** to show the inheritance pattern of the disorder.

Figure 3

		Paul's alleles	
		....	....
Eilidh's alleles	F	FF	....
	....	Ff	....

My family tree is just full of surprises...

...I never knew that Auntie Joan was a world-class belly dancer, or that Great Uncle Simon swam the English Channel. Anyway, make sure you understand how disorders can be inherited and how family trees show inheritance.

☐  ☐  ☐

Variation and Evolution

Excellent, some variation in these pages at last — oh wait, there are still questions for you to answer...

Warm-Up

Organisms of the same species have differences.
These differences can be caused by genes or by the environment.

Differences in genes are caused by mutations.
Complete the sentences below by circling the correct word or words from each pair.

A mutation is a **random / non-random** change in an organism's DNA.

Mutations occur **very rarely / continuously**.

Mutations **always / very rarely** lead to a new phenotype.

Variation is what allows species to evolve. All of today's species have evolved from simple life forms. Species that aren't able to evolve in response to changes in their environment will become extinct (die out).

When did simple life forms first start to develop? Tick one box.

- [ ] **A** Less than three thousand years ago.
- [ ] **B** Around three million years ago.
- [ ] **C** More than three billion years ago.

Q1 There are many reasons why a species might become extinct. Draw lines to match each reason for extinction on the left with **one** correct example on the right.

A catastrophic event kills every member of the species.	A new fungal pathogen is accidentally introduced to a habitat. It kills every member of a species of toad.
The environment changes too quickly.	An island's rainforest is completely chopped down. All of the striped monkey's habitat is destroyed.
A new disease kills every member of the species.	A rare plant species only lives on one volcano. They are all wiped out when the volcano erupts.

Q2 Do you think variation in the characteristics below is mainly due to genes, the environment, or a combination of both? Tick **one** box for each characteristic.

	Genes	Environment	Both
a) Language(s) spoken	☐	☐	☐
b) Eye colour	☐	☐	☐
c) Naturally curly hair	☐	☐	☐
d) Scars	☐	☐	☐
e) Intelligence	☐	☐	☐

Q3 Mr O'Riley breeds racehorses. He bred his best male racehorse, Snowball, with his best female racehorse, Goldie. Snowball and Goldie are both black. They produced a black foal, Cinnamon.

a) Cinnamon did not become a champion racer.
Which of the following is a possible explanation for this? Tick **one** box.

☐ **A** Cinnamon did not inherit any DNA from Snowball and Goldie.
☐ **B** Cinnamon was not raised in an environment that made him good at racing.
☐ **C** Genes do not affect racing ability.

b) Is Cinnamon's colour due to genes or to the environment?

....

Q4 **Figure 1** shows a buff-tip moth. The buff-tip moth looks very similar to a broken stick. This makes it very difficult for predators to see buff-tip moths. The statements below describe how this feature might have evolved.

Write each of the numbers **1** to **4** in the boxes below to show the order the statements should be in. The first has been done for you.

[1] In a moth population, individuals showed variation in their appearance. Some had genes that made them look a bit like a stick.
☐ So the stick-like moths were more likely to survive and reproduce.
☐ Genes that made the moths look like sticks were more likely to be passed on to their offspring.
☐ Birds were less likely to spot and eat the moths that blended in with their environment. Birds were more likely to eat the moths that stood out like a sore thumb.

Figure 1

If you find these jokes funny, you've got a highly evolved sense of humour...

Variation and evolution are two ideas that go hand in hand — a species simply cannot evolve if all the individuals are the same. There's always at least some variation in a population though — mutations make sure of that.

Antibiotic-Resistant Bacteria

Bacteria can evolve really quickly. This is making it more difficult to treat infections.

Warm-Up

Antibiotics are drugs that kill bacteria. Unfortunately for us, some bacteria have evolved to become antibiotic-resistant. This means they are no longer killed by an antibiotic.

Which of the following rules could help reduce the spread of antibiotic resistance? Tick two boxes.

- [ ] Farmers should always give farm animals antibiotics when they're ill.
- [ ] Doctors should only prescribe antibiotics when they really need to.
- [ ] Patients should take all the antibiotics a doctor prescribes for them.
- [ ] Doctors should use antibiotics to treat patients infected with viruses.

PRACTICAL

Q1 Marie was investigating how antibiotic resistance can evolve in bacteria. She added an antibiotic to a plate. The plate contained the nutrients needed for bacteria to grow. Marie then introduced a population of bacteria to the plate. The population included antibiotic-resistant bacteria and non-resistant bacteria. She left the plate for four days to allow the bacteria to grow (see **Figure 1**).

Figure 1

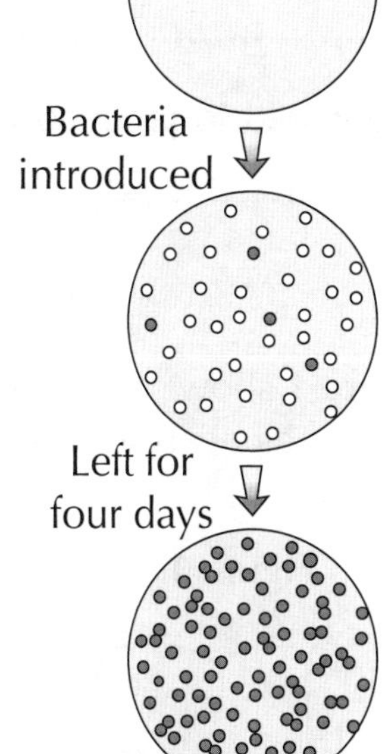

Key:
● Antibiotic-resistant bacteria
○ Non-resistant bacteria

a) Suggest **one** possible reason why the number of non-resistant bacteria decreased after four days.

....

b) Describe **one** other way that the bacterial population changed after four days.

....

....

c) What evidence is there to suggest that the bacterial population has evolved? Tick **one** box.

- [ ] **A** The bacteria have formed a new species.
- [ ] **B** The bacteria have stopped mutating.
- [ ] **C** There has been a change in the inherited characteristics of the population.

I'm feeling anti-biology — I'm resistant to any learning...

Bacteria evolve so quickly that we can see it happening with our own eyes — this is why the evolution of antibiotic resistance in bacteria provides really strong evidence to support the theory of natural selection.

Selective Breeding and Genetic Engineering

Hold onto your hats... things are about to get interesting. Sort of.

Warm-Up

Selective breeding is where humans choose which plants or animals are going to breed. Individuals are chosen based on their characteristics.

How long has selective breeding been happening? Tick one box.

- ☐ A Since the discovery of genes in 1866.
- ☐ B Since the 1990s.
- ☐ C For millions of years.
- ☐ D For thousands of years, since wild plants and animals were first domesticated.

Genetic engineering involves scientists making changes to organisms' DNA.

Choose some of the words from the list to fill in the gaps in the passage below.

characteristic   insulin   genome   gene

Genetic engineering involves transferring a gene from one organism's .... to another's. The gene codes for a desired .... Bacteria have been genetically engineered to produce .... to treat diabetes.

Q1 **Figure 1** shows a modern domesticated carrot and a wild carrot. The modern carrot has been selectively bred from the wild carrot over many years.

Figure 1

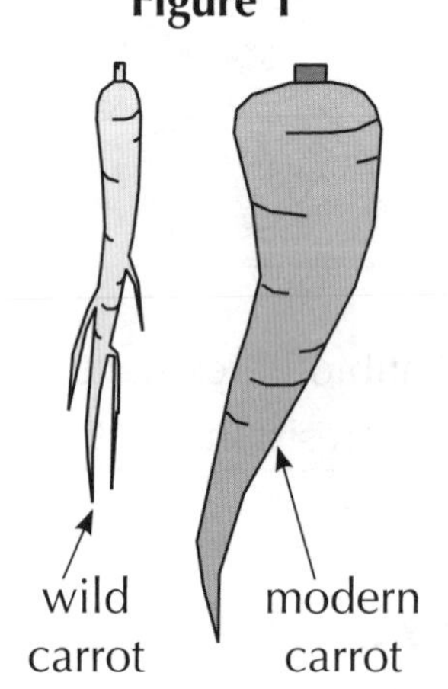

a) The characteristics of a modern carrot are very different to those of a wild carrot. Write down **two** ways in which the modern carrot in **Figure 1** is different to the wild carrot.

1. ....

2. ....

b) Suggest **one** reason why people may have bred the modern carrot to have the characteristics that it does.

....

....

c) Leaf blight is a disease that can cause damage to modern carrots. Some carrot plants are more badly affected by leaf blight than others. A crop breeder wants to produce a variety of carrot plant with a high resistance to leaf blight. Describe how the breeder could do this.

....

....

....

....

Q2 Herbicides are chemicals that kill plants. They are sprayed on fields to kill weeds. Maize plants can be genetically modified to make them resistant to herbicides.

a) Why it would be beneficial to a farmer if her maize plants were resistant to herbicides? Tick **one** box.

☐ **A** The farmer could spray herbicides on fields to kill weeds without damaging the maize plants.

☐ **B** The farmer wouldn't need to use herbicides to kill weeds any more.

☐ **C** The maize plants would produce their own herbicides.

b) **Table 1** shows the average yield produced by two types of maize crop (**A** and **B**). One type is herbicide-resistant, but the other is not. Which maize crop (**A** or **B**) is most likely to be herbicide-resistant? Give a reason for your answer.

Table 1

Maize crop	Yield (kg per m^2)
A	1.23
B	1.09

....

....

c) Some people are worried that genetically modified (GM) crops could pollinate wild plants. If this happened with GM maize, it is possible that the genes for herbicide resistance would be passed on to wild plants.

i) Suggest why it could be a big problem if the genes for herbicide resistance were passed on to weeds.

....

....

ii) Suggest **one** other reason why people might be worried about growing GM maize.

....

....

Could they genetically engineer broccoli to taste like chocolate...?

Selective breeding and genetic engineering have the same goal — they're both used to change the characteristics of organisms so that they are more beneficial to humans. However, remember the methods used are very different.

Fossils

Antibiotic-resistant bacteria are one source of evidence for evolution. Fossils are another.

Warm-Up

<u>Fossils</u> are the remains of organisms from many thousands or millions of years ago. Scientists can use <u>fossils</u> to study what life on Earth used to be like.

Where are most fossils found? Tick <u>one</u> box.

in soil ☐ in rock ☐ in water ☐ in trees ☐

Q1 The bodies of organisms are not usually preserved. This is because they are decayed by microbes.

a) A preserved body of a woolly mammoth was found in the Arctic permafrost (permanently frozen ground). Suggest why it has not been decayed by microbes.

....

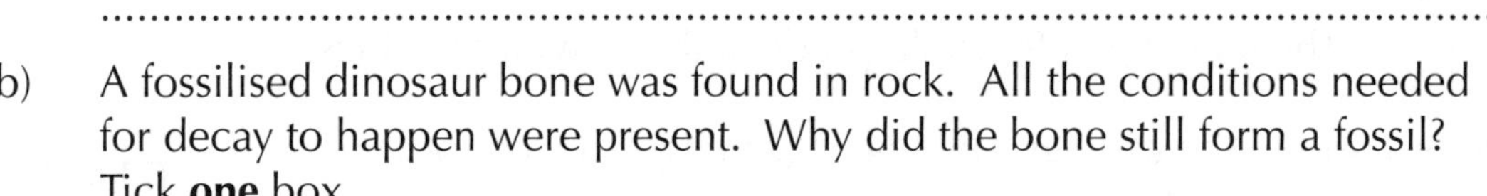

b) A fossilised dinosaur bone was found in rock. All the conditions needed for decay to happen were present. Why did the bone still form a fossil? Tick **one** box.

☐ **A** The bone was replaced by minerals as it decayed.

☐ **B** Microbes can't decay bones.

☐ **C** The bone left an impression in the rock.

Q2 **Figure 1** shows a cliff face and the places where three fossils were found in the cliff.

a) The most recent fossil is of a fossilised early fish. Which of the three fossils (**X**, **Y** or **Z**) found in the cliff is most likely to be the fish fossil?

....

Figure 1

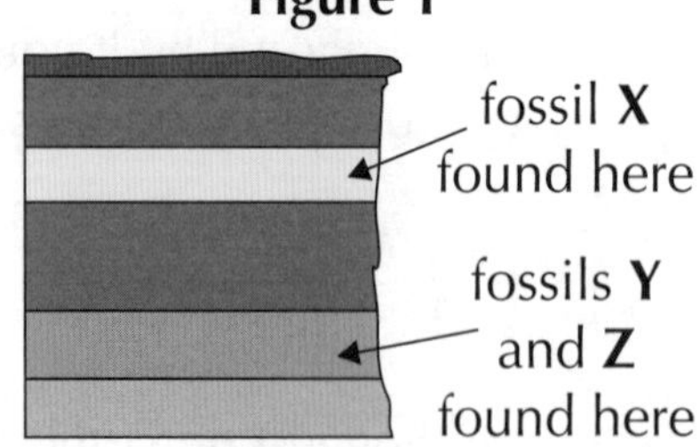

b) Fossils **X**, **Y** and **Z** are all animal fossils. Suggest why animal fossils are more common than plant fossils.

....

....

Think about how the bodies of animals and plants are different.

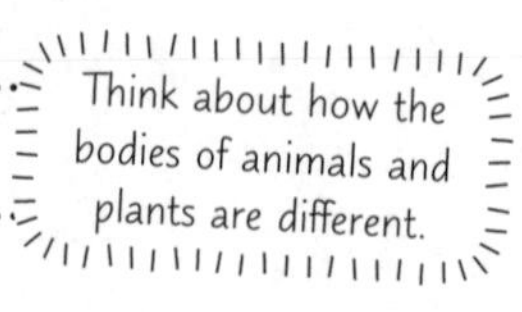

I thought I'd found a dinosaur skull — turns out it was a fossil arm...

Fossils are really useful — they can show us how organisms have changed over millions over years. However, they don't tell us anything about how life on Earth began — any fossils that formed at that time may have since been destroyed.

Classification

Biologists love to keep things neat and tidy by sorting organisms into different groups.

Warm-Up

The Linnaean system is used to classify living organisms based on how closely related they are. In the Linnaean system, kingdoms are the highest level of classification.

Fill in the missing labels on the diagram to show the names of each level of classification.

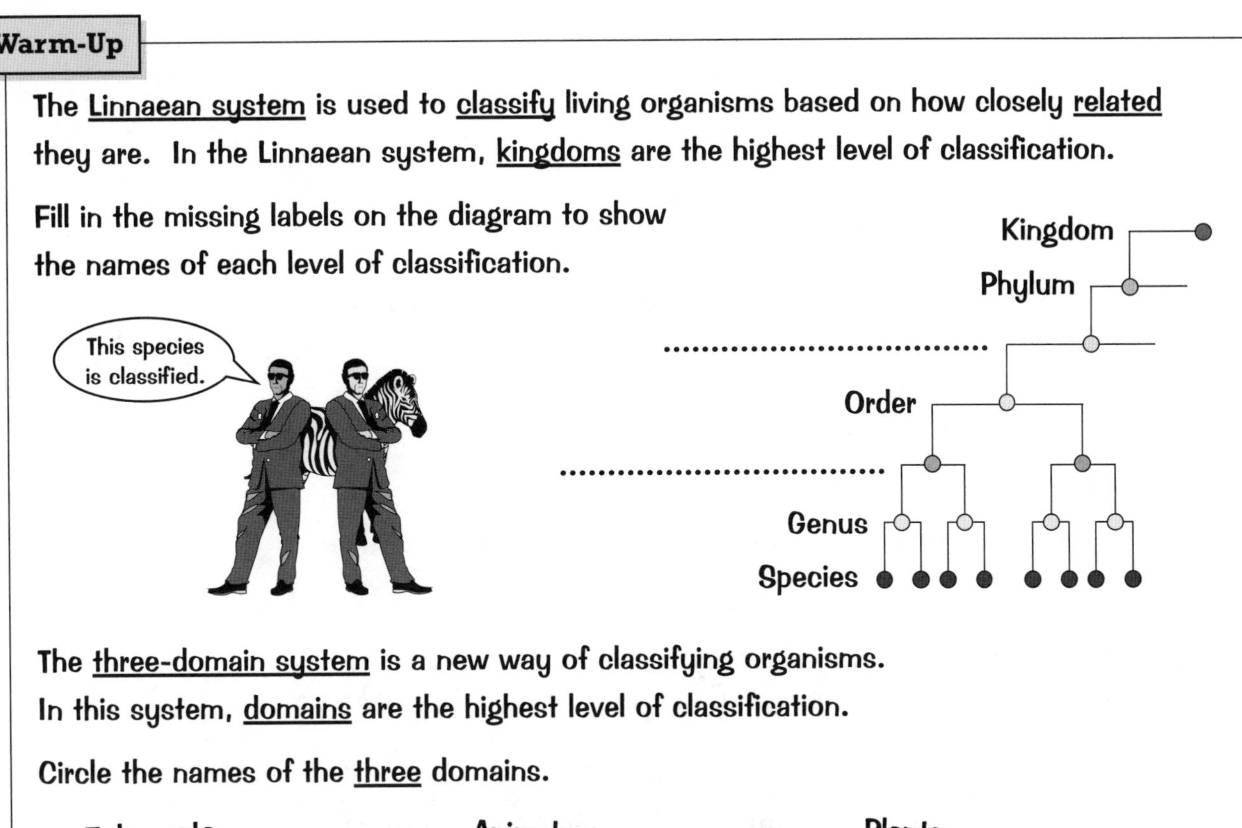

The three-domain system is a new way of classifying organisms. In this system, domains are the highest level of classification.

Circle the names of the three domains.

Eukaryota Fungi Animals Bacteria Plants Archaea

Q1 Three duck species are shown in **Figure 1**. Their common names are shown in **bold** and their binomial names are shown in *italic*.

Figure 1

Mallard
Anas platyrhynchos

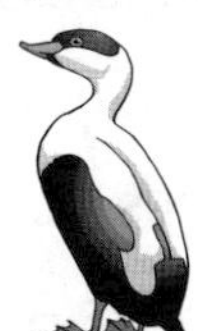

Eider
Somateria mollissima

Teal
Anas crecca

a) Which genus does the eider belong to? Circle the correct answer below.

Anas *Somateria* *mollissima* *crecca*

b) What is the species name of the teal?

....

c) Which **two** species belong to the same genus? Write their common names.

1. .... 2. ....

d) Name the domain that all three species belong to.

....

Q2 **Figure 2** shows how four different groups of organisms are related. Three ancestors (**X**, **Y** and **Z**) of these groups are also shown.

Figure 2

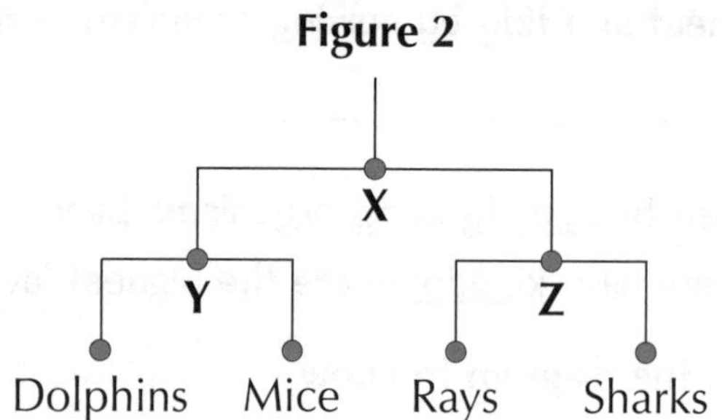

Tick the boxes to show whether the statements below are **true** or **false**.

		True	False
a)	**X** lived longer ago than **Z**.	☐	☐
b)	**Y** is an ancestor of rays.	☐	☐
c)	**X** is a common ancestor of dolphins and sharks.	☐	☐
d)	**Y** is the most recent common ancestor of dolphins and mice.	☐	☐

Q3 **Table 1** shows the names of the family, genus and species that four species of beetle are classified into.

Table 1

Species	Family	Genus	Species name
A	Carabidae	*Carabus*	*violaceus*
B	Carabidae	*Amara*	*aenea*
C	Silphidae	*Silpha*	*atrata*
D	Carabidae	*Carabus*	*nemoralis*

a) Which species (**B**, **C** or **D**)...

i) ...has the same genus name as species **A**? ....

ii) ...is **not** in the same family as species **A**? ....

iii)...is most closely related to species **A**? ....

iv)...is most distantly related to species **A**? ....

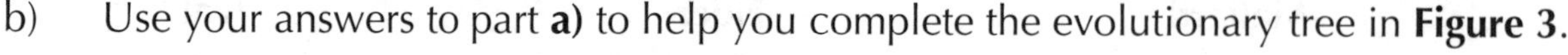

b) Use your answers to part **a)** to help you complete the evolutionary tree in **Figure 3**.

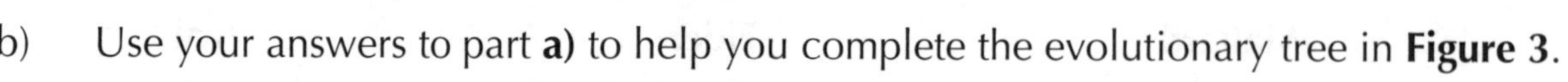

Figure 3

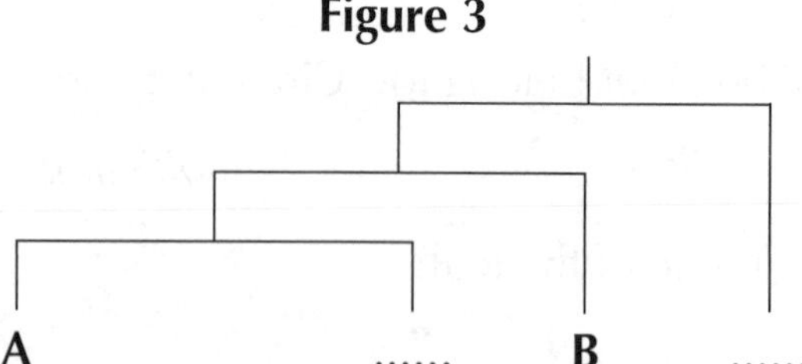

Sort organisms into different folders — that's how you phylum...

Remember that the new three-domain system isn't a replacement for the Linnaean system. Organisms are still sorted into a kingdom, phylum, class, etc. But there's now an extra layer of classification at the top — the three domains.

Competition, Abiotic and Biotic Factors

Congratulations, you've won our competition. Your prize — a set of ecology questions to answer.

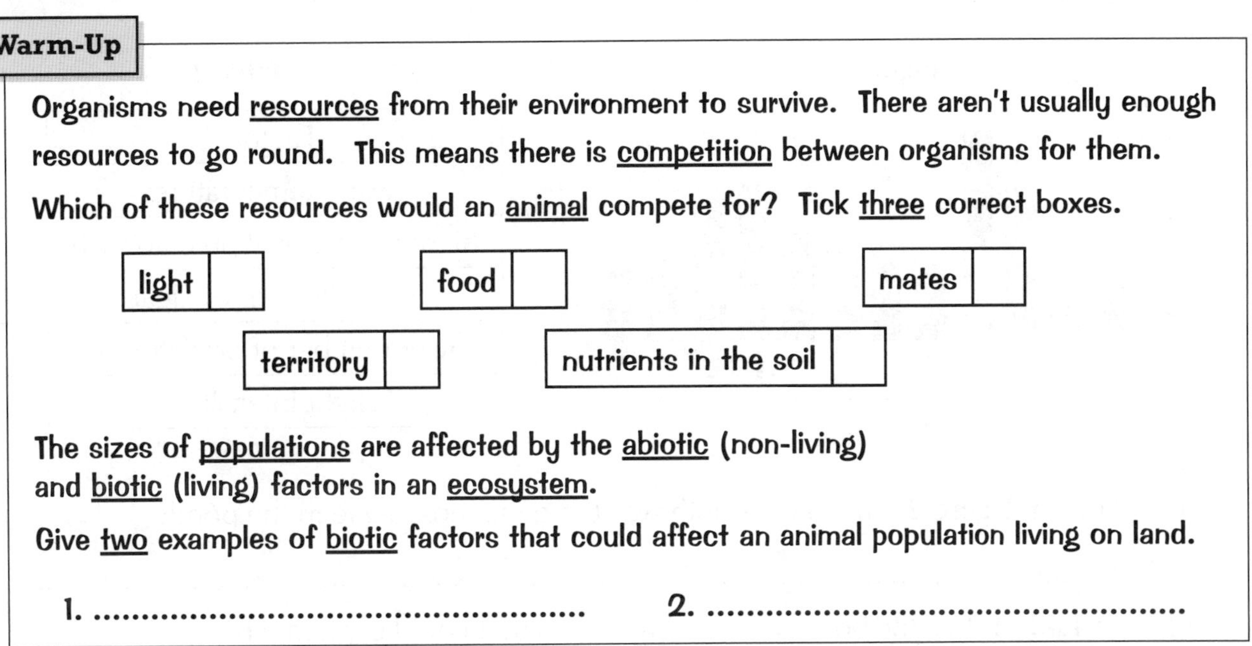

Warm-Up

Organisms need <u>resources</u> from their environment to survive. There aren't usually enough resources to go round. This means there is <u>competition</u> between organisms for them.

Which of these resources would an <u>animal</u> compete for? Tick <u>three</u> correct boxes.

light ☐ food ☐ mates ☐

territory ☐ nutrients in the soil ☐

The sizes of <u>populations</u> are affected by the <u>abiotic</u> (non-living) and <u>biotic</u> (living) factors in an <u>ecosystem</u>.

Give <u>two</u> examples of <u>biotic</u> factors that could affect an animal population living on land.

1. .... 2. ....

Q1 **Figure 1** shows a food web for an ecosystem on a small island. Snakes have recently arrived on the island. This has caused a decrease in the population of lizards.

Figure 1

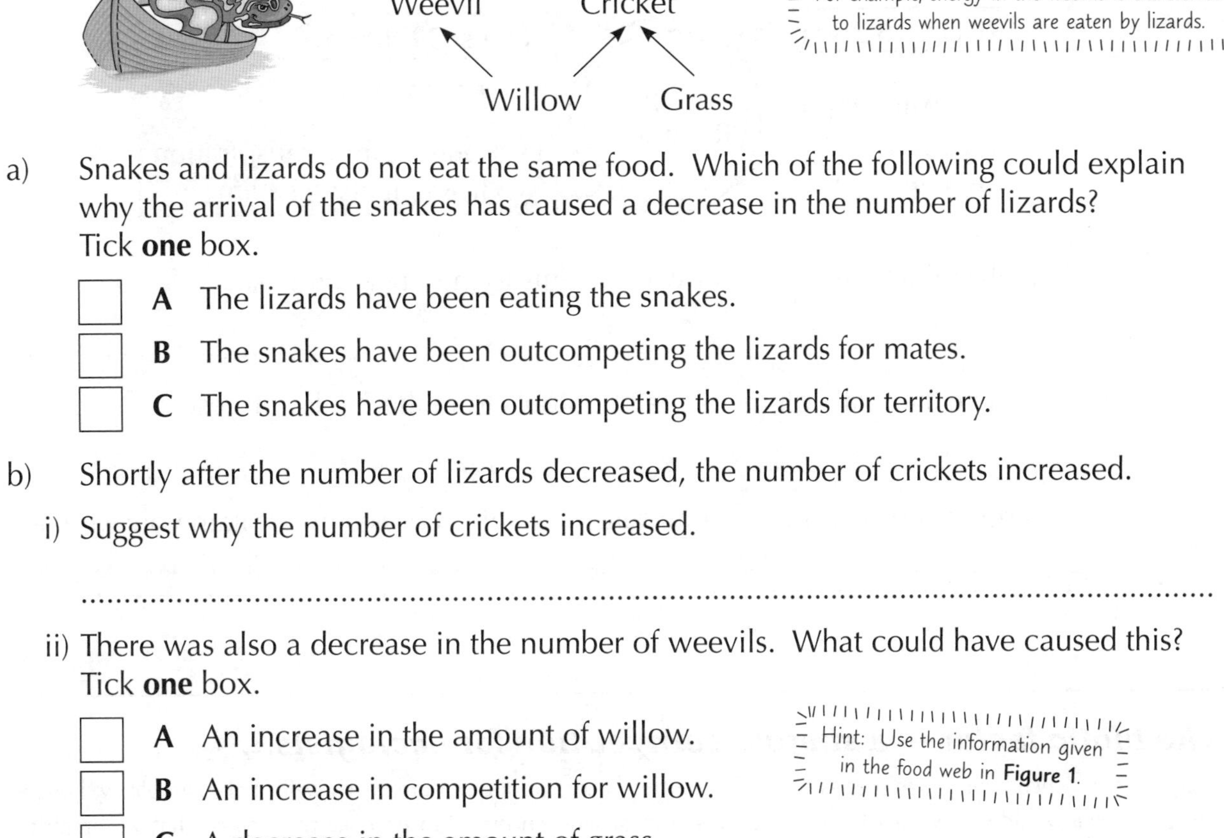

The arrows on a food web show the direction of energy flow as one organism eats another. For example, energy in the weevils is transferred to lizards when weevils are eaten by lizards.

a) Snakes and lizards do not eat the same food. Which of the following could explain why the arrival of the snakes has caused a decrease in the number of lizards? Tick **one** box.

☐ **A** The lizards have been eating the snakes.

☐ **B** The snakes have been outcompeting the lizards for mates.

☐ **C** The snakes have been outcompeting the lizards for territory.

b) Shortly after the number of lizards decreased, the number of crickets increased.

i) Suggest why the number of crickets increased.

....

ii) There was also a decrease in the number of weevils. What could have caused this? Tick **one** box.

☐ **A** An increase in the amount of willow.

☐ **B** An increase in competition for willow.

☐ **C** A decrease in the amount of grass.

Hint: Use the information given in the food web in **Figure 1**.

Q2 Algae are tiny organisms that are eaten by fish. Like plants, algae carry out photosynthesis. **Figure 2** shows how the size of a population of algae in a pond varied throughout one year. **Table 1** shows some factors that might affect the size of the algae population.

Figure 2

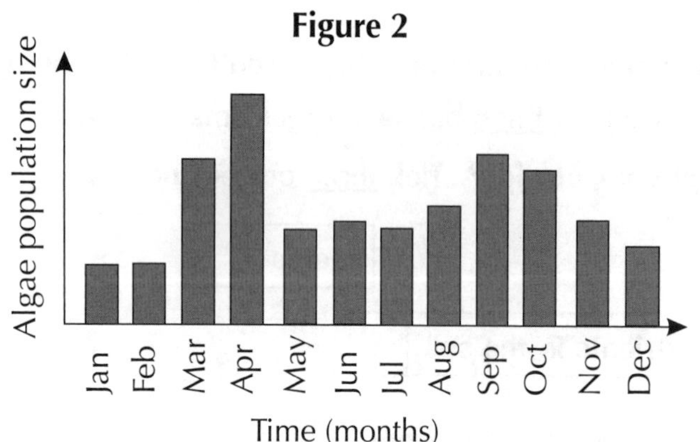

Table 1

Factor	
water temperature	
higher number of predator fish	
more diseases of algae	
lower number of predator fish	
light intensity	

a) Look at **Figure 2**. In which month was there the most algae in the pond?

....

b) In **Table 1**, tick the boxes next to any factors that could explain why there were more algae in the pond during this month.

Q3 **Figure 3** shows the volume of water that goes in and out of the Zig Rainforest (on the planet Zog) per day.

Figure 3

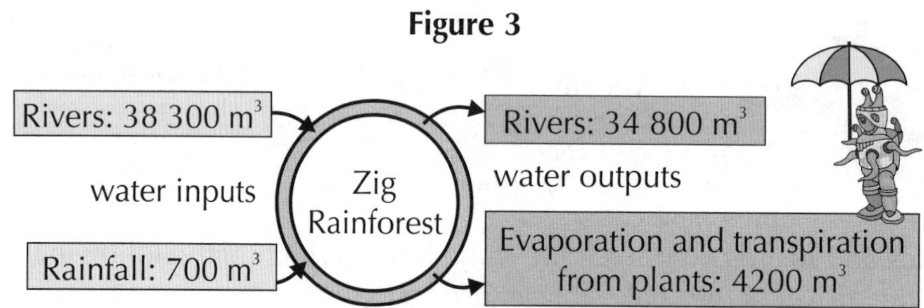

a) i) Calculate the total volume of water inputs to the Zig Rainforest.

.... m^3

ii) Calculate the total volume of water outputs from the Zig Rainforest.

.... m^3

b) In terms of water, is the Zig Rainforest a stable ecosystem? Explain your answer.

....

The biotic factor — a singing competition for ecologists...

Every species depends on other species to survive — this is true about us humans as well. For example, without animals and plants, we wouldn't have any food to eat. We also depend on plants to produce the oxygen that we breathe.

Adaptations

If you've ever wondered why tigers are stripy, or why birds migrate, it's all to do with adaptations.

Warm-Up

Adaptations are the features of organisms that allow them to survive in their natural habitat. Adaptations can be structural, behavioural or functional.

Draw lines to match each type of adaptation to the correct description.

Behavioural adaptation	An adaptation that is a feature of an organism's body parts.
Functional adaptation	A way that an organism is adapted to act.
Structural adaptation	An adaptation related to things that go on inside an organism's body.

Q1 Tick the boxes to show whether the following adaptations are structural, behavioural or functional. The first one has been done for you

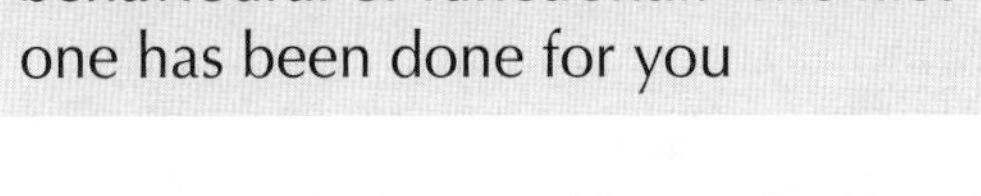

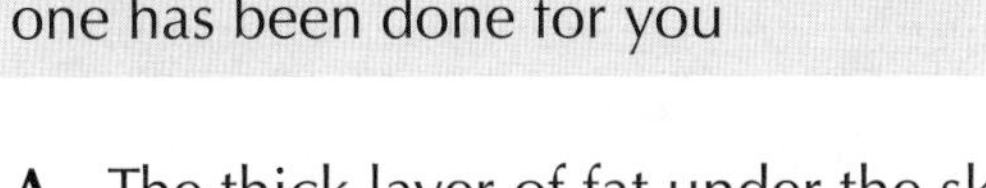

	Structural Adaptation	Behavioural Adaptation	Functional Adaptation
A The thick layer of fat under the skin that keeps a walrus warm.	☑	☐	☐
B A hedgehog slowing its metabolism to save energy in winter.	☐	☐	☐
C A bird flying to a warmer area in winter.	☐	☐	☐
D The stripes on a tiger that help it to hide from prey in long grass.	☐	☐	☐

Q2 **Figure 1** shows a polar bear and a small rodent called a long-eared jerboa.

Figure 1

Diagrams are not to scale.

a) The shape of the long-eared jerboa gives it a large surface area compared to its volume. Suggest **one** way in which the shape of the long-eared jerboa gives it a large surface area compared to its volume.

....

b) Choose the correct way to complete the sentence below. Tick **one** box.

Having a large surface area compared to its volume...

☐ **A** ...helps an organism to lose heat.

☐ **B** ...makes it difficult for an organism to lose heat.

☐ **C** ...has no effect on an organism's ability to lost heat.

c) Polar bears live in the Arctic, while long-eared jerboas are found in hot deserts. Suggest how the shape of the long-eared jerboa is an adaptation to its environment.

Q3 Scientists are studying the adaptations of two species of shrimp. Species **A** lives around deep sea vents. These vents are areas on the sea bed in very deep water, which release hot liquid. Species **B** lives in shallow water near the coast.

a) Suggest how the following conditions are different around deep sea vents compared to shallow coastal waters.

i) light intensity

ii) water pressure

b) No photosynthesising organisms are found around deep sea vents. Suggest why.

c) Suggest which shrimp species (**A** or **B**) is an extremophile. Give a reason for your answer.

d) The scientists have found that Species **B** is unable to survive the conditions around deep sea vents. Suggest why Species **A** can survive around deep sea vents, but Species **B** cannot.

I'm well-suited to my natural environment — on a sofa, watching TV...

You need to be able to explain how a particular adaptation helps an organism to survive. With these types of questions, you might have to work it out from the information given to you — so make sure you read the question carefully.

Food Chains

Food chains — part of every ecologist's diet. Here are a few questions to keep you well-fed.

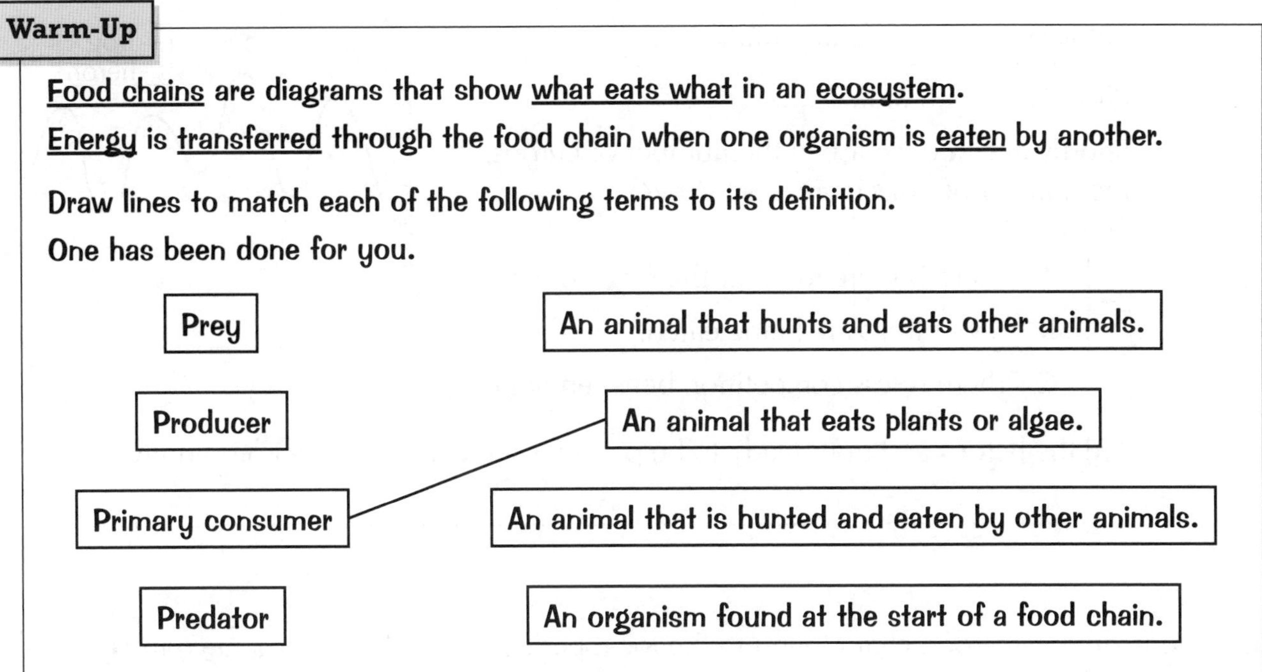

Warm-Up
Food chains are diagrams that show what eats what in an ecosystem.
Energy is transferred through the food chain when one organism is eaten by another.
Draw lines to match each of the following terms to its definition.
One has been done for you.
Prey
Producer
Primary consumer
Predator
An animal that hunts and eats other animals.
An animal that eats plants or algae.
An animal that is hunted and eaten by other animals.
An organism found at the start of a food chain.

Q1 **Figure 1** shows a food chain found on the beach.

Figure 1

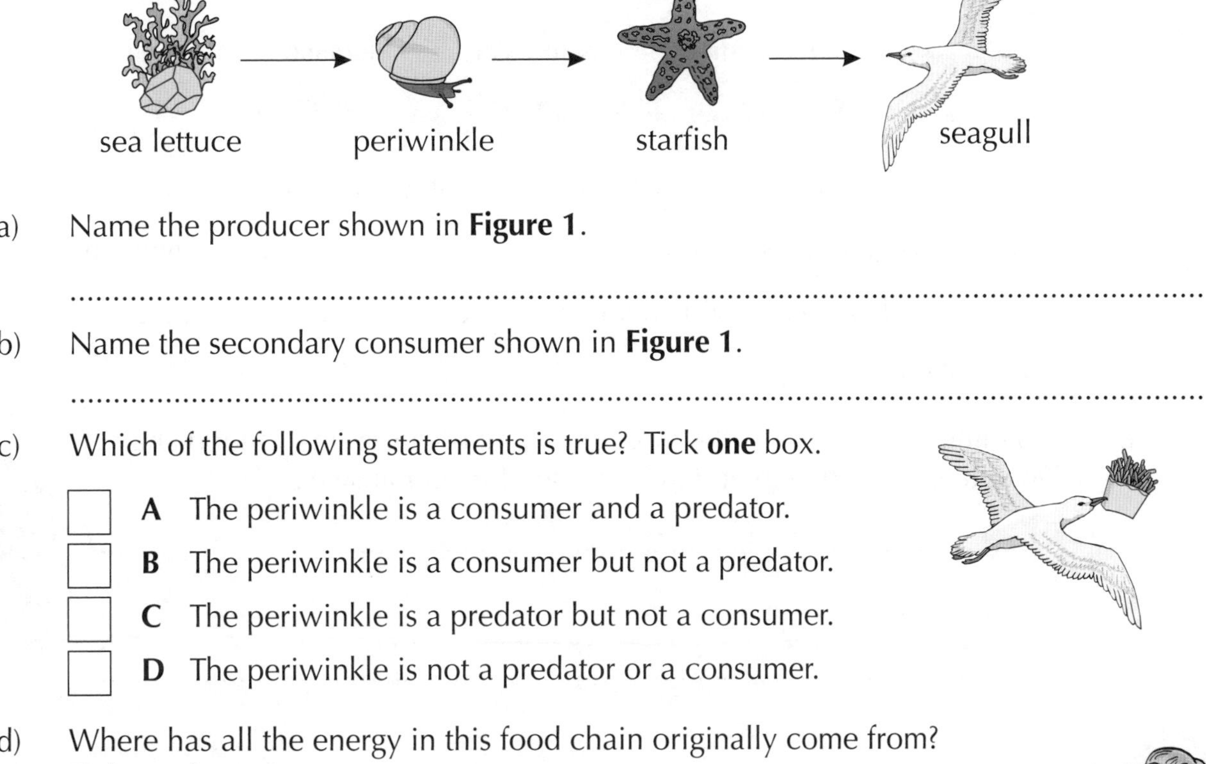

a) Name the producer shown in **Figure 1**.

....

b) Name the secondary consumer shown in **Figure 1**.

....

c) Which of the following statements is true? Tick **one** box.

- [ ] **A** The periwinkle is a consumer and a predator.
- [ ] **B** The periwinkle is a consumer but not a predator.
- [ ] **C** The periwinkle is a predator but not a consumer.
- [ ] **D** The periwinkle is not a predator or a consumer.

d) Where has all the energy in this food chain originally come from? Tick **one** box.

- [ ] **A** seawater
- [ ] **B** the sea lettuce
- [ ] **C** the Sun

Q2 **Figure 2** shows a population graph for herons and frogs. The number of herons and frogs rise and fall in cycles.

a) Which animal is the predator and which is the prey?

Predator: ....

Prey: ....

Figure 2

Number of animals

Time

— frog
— heron

b) Shortly after a decrease in the number of herons, the number of frogs increases. Why?
Tick **one** box.

☐ **A** There is more food for the frogs to eat.

☐ **B** Fewer frogs are being eaten.

☐ **C** There is less competition between the frogs.

c) Are the herons and frogs likely to be part of a stable or unstable community?

....

Q3 Mo is studying a food chain found in the local park. The food chain is shown in **Figure 3**.

Figure 3

Oak tree ⟶ Caterpillar ⟶ Blackbird ⟶ Fox

a) Complete the passage below about the food chain in **Figure 3** using the correct words from the box.

respiration	caterpillar	biomass
oak tree	photosynthesis	blackbird

The .... makes glucose using ....

Some of this glucose is used to produce .... (living material).

This is passed along the food chain when the ....

feeds on the oak tree.

b) Mo thinks the oak tree is the most important organism in this food chain. Do you agree or disagree? Give a reason for your answer.

....

....

....

The heron and the frog — a fairy tale with a very different ending...

Food chains aren't just about things eating other things. They help us to understand how biomass is transferred through an ecosystem. They can also help us to predict how a change in the population of one species could affect other species.

Using Quadrats and Transects

PRACTICAL

Ecology practicals are the best practicals. You get to go outside, get some fresh air and get up close and personal with some nature. Best of all, there's not a test tube in sight.

Warm-Up

Quadrats and transects are methods of studying the abundance (population size) and distribution of species in a habitat.

Label the diagrams below to show which is a quadrat and which is a transect.

....

Which of the following can quadrats be used for? Tick two boxes.

- [ ] **A** To estimate the total population size of a species in an area.
- [ ] **B** To investigate how temperature changes across an area.
- [ ] **C** To count exactly how many organisms live in a large area.
- [ ] **D** To collect data along a transect.

Q1 Eve and Will work together to study the distribution of harebells in a meadow. They each pick a sample area within the meadow. They then use a quadrat to estimate the population size of harebells in their sample area.

a) If Eve and Will are studying the 'distribution of harebells in a meadow', what are they hoping to find out? Tick **one** box.

- [ ] **A** How many harebells there are in the entire meadow.
- [ ] **B** Where the harebells are found in the meadow.

b) Eve and Will each estimated the mean number of harebells per m^2 in their sample area. The sentences below describe the method they used. Put them in the correct order by numbering the boxes from **1** to **4**. The first one has been done for you.

- [ ] Divide the total number of harebells by the total number of quadrats to find the mean.
- [ ] Count all the harebells within the quadrat.
- [1] Put the quadrat on the ground at a random point in the sample area.
- [ ] Repeat the previous steps at several other random points.

Q2 Azi wanted to estimate the number of gnomes in a garden. He placed a 0.5 m^2 quadrat down at five random points and counted the number of gnomes in each quadrat. He recorded his results in **Table 1**.

Table 1

Quadrat number	1	2	3	4	5
Number of gnomes	3	1	2	4	0

a) What is the median number of gnomes per 0.5 m^2 quadrat?

The median is the middle value when the data is put in order.

.... gnomes per quadrat

b) i) Calculate the mean number of gnomes per 0.5 m^2 quadrat.

.... gnomes per quadrat

ii) What is the mean number of gnomes in a 1 m^2 area?

.... gnomes per 1 m^2

c) The total area of the garden is 160 m^2.
Use your answer to **b) ii)** to estimate the total number of gnomes in the whole garden.

.... gnomes

Q3 Sandy uses a transect to investigate buttercup distribution from the middle of a field to a pond.

a) On **Figure 1**, draw **one** way that Sandy could set up her transect.

Figure 1

MIDDLE OF FIELD

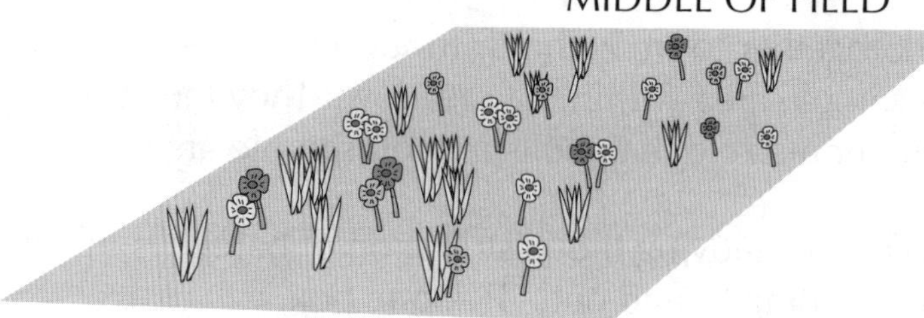

POND

b) Sandy sets up a second transect at a different location in the field.
She then collects data from this transect in the same way she did for the first transect.
Explain why Sandy carried out a second transect.

....

....

Gosh, that's a lot of gnomes for one garden...

It's not always the number of organisms in a quadrat that's recorded — you can record the percentage cover instead. This is the percentage area of the quadrat covered by a species. It's useful when it's difficult to count individual organisms.

The Water Cycle and The Carbon Cycle

Don't worry, you haven't accidentally opened a geography book — the water cycle and the carbon cycle are both important for the survival of ecosystems.

Warm-Up

The water cycle recycles water.

The diagram on the right shows part of the water cycle.

Write down the letter in the diagram that represents each of these processes:

Evaporation .... Precipitation ....

The carbon cycle recycles carbon in an ecosystem. Carbon moves through food chains and is cycled back to the air to be used by plants.

What is the most common form of carbon found in the air?

....

Q1 **Figure 1** shows a large island. A mountain is located at the centre of the island. Air is pushed upwards when it hits the mountain.

Figure 1

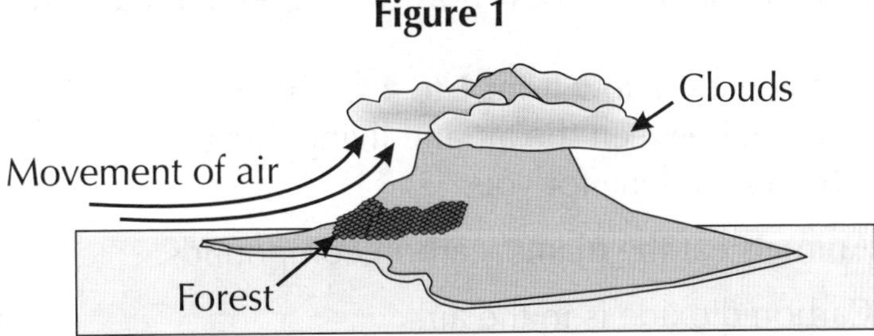

a) The passage below describes how clouds form over the island. Complete the passage by circling the correct word in each pair.

Water **evaporates / condenses** from the island and the sea. This water enters the air as water vapour. Transpiration from the **animals / plants** on the island also releases water vapour. Clouds form over the mountain when the water vapour in the air **condenses / precipitates**.

b) There is much more cloud above the mountain than there is over the rest of the island.

i) What must happen to water vapour in order for it to form clouds? Tick **one** box.

☐ It must warm up. ☐ It must cool down.

ii) Using your answer to **b) i)**, suggest why lots of clouds form above the mountain.

....

....

....

Look back at **Figure 1** and the information given at the start of the question about the movement of the air.

c) Give **one** reason why the island's forest needs the water cycle to survive.

....

Q2 **Figure 2** shows part of the carbon cycle.

Figure 2

CO_2 in the air

plant   animal

a) Add an arrow or arrows labelled **P** to **Figure 2** to represent photosynthesis.

b) Add an arrow or arrows labelled **R** to **Figure 2** to represent respiration.

c) Add an arrow or arrows labelled **F** to **Figure 2** to represent the movement of carbon compounds through feeding.

Q3 Answer the following questions to show how the stages in the carbon cycle are ordered.

a) Number the sentences below from **1** to **4** to show how carbon moves between the air and living things. The first one has been done for you.

.... Animals eat the plants' carbon compounds.

......1...... Carbon dioxide is in the air.

.... Plants and animals die.

.... Plants take in carbon dioxide for photosynthesis and make carbon compounds.

You first, then you. That's an order.

b) Write a point 5 to complete the cycle and show how carbon is returned to the air.

Point 5: ....

....

The water cycle makes my head spin — now I know how my laundry feels...

You've learned about the water cycle and the carbon cycle, but they're not the only things that need to be recycled for an ecosystem to be stable — every element used by living things needs to be returned to the ecosystem so it can be reused.

Biodiversity, Waste and Global Warming

Earth is home to a huge number of species. Unfortunately, we're causing biodiversity to decrease.

Warm-Up

Biodiversity is the variety of different species on Earth or in an ecosystem.
Human activities such as waste production and pollution are reducing biodiversity.

Draw lines to show whether the forms of pollution on the left are most likely to pollute land, rivers and streams or air.

Smoke	Land
Household waste sent to landfill	Rivers and streams
Sewage	Air

Another threat to biodiversity is global warming. This is caused by increasing levels of greenhouse gases, such as carbon dioxide. Name one other greenhouse gas.

....

Q1 One way to measure a person's impact on the Earth is to calculate an ecological footprint. The more waste a person produces and the more energy they use, the bigger their ecological footprint.

Two men, John and Derek, calculate their ecological footprints. They find that John has a bigger ecological footprint than Derek has.

a) Which of the following are possible reasons for John's bigger ecological footprint? Tick **two** boxes.

- [ ] John buys more belongings, which use more resources to manufacture.
- [ ] John recycles more than Derek.
- [ ] John drives a car and Derek rides a bicycle.

b) Suggest **one** thing John could do to reduce the size of his ecological footprint.

....

c) People who live in developed countries often have a higher standard of living than people who live in developing countries.
Explain why people living in developed countries are likely to have a bigger ecological footprint than people living in developing countries.

....

....

....

Q2 **Table 1** shows data for two lakes (**A** and **B**).

Table 1

	Lake **A**	Lake **B**
Number of different fish species	15	2
Number of other species	75	4

a) i) How many more species does lake **A** have than lake **B**?

.... more species

ii) Which lake (**A** or **B**) has the greatest biodiversity? Justify your answer.

....

....

b) Perch are a type of fish. In lake **A**, perch feed on minnows, sticklebacks and roach. This is shown in **Figure 1**. The perch in lake **B** can only feed on minnows. This is shown in **Figure 2**.

Figure 1

Perch

Stickleback → Perch, Minnow → Perch, Roach → Perch

Figure 2

Perch

Minnow → Perch

A disease wipes out most of the minnows in both lakes.
It does not affect any other species.

i) A decrease in the number of minnows is expected to cause a decrease in the number of perch. Suggest why.

....

ii) Which perch population would you expect to be affected more by the minnow disease — those in lake **A** or those in lake **B**?

....

iii) How does the level of biodiversity in a lake affect the likelihood of a perch population being stable?

A population is stable if it doesn't change much over time.

....

c) Suggest **one** way that global warming could reduce biodiversity in a lake.

....

I make sure nothing goes to waste — all my jokes are recycled...

Biodiversity is decreasing and unfortunately there's no easy fix. We need land to produce enough food for us all. We also need resources to keep our standard of living. All that waste we produce has to go somewhere too.

Land Use and Maintaining Biodiversity

Some more doom and gloom I'm afraid — the way we use land is causing problems for biodiversity.

Warm-Up

Lots of land is cleared to make room for human activities.
These activities include building, farming, quarrying and dumping waste.
Many species lose their habitat as a result. This affects biodiversity.

Tick the boxes to show whether the following statements are true or false.

	True	False
Burning peat releases carbon dioxide.	☐	☐
Recycling increases the amount of land needed for landfill.	☐	☐
No animals or plants are found living on peat bogs.	☐	☐
Some governments have made rules about how much deforestation can happen.	☐	☐

Q1 A rare species of monkey is found in only one area of tropical forest. The forest is under threat from deforestation. Scientists say the monkey's habitat should be made into a nature reserve. This would stop the habitat being destroyed.

a) Which of the following words could be used to describe this species of monkey? Tick **one** box.

☐ **A** extinct ☐ **B** endangered ☐ **C** common

b) Suggest **one** reason why people might want to cut down the forest.

....

c) Protecting the forest from deforestation will help to protect the monkeys. Suggest **one** other benefit of protecting the forest.

....

d) Describe **one** way that scientists could stop the monkeys from dying out if the forest is cut down.

....

....

e) Some of the monkeys' habitat has already been destroyed. The scientists want these areas of habitat to be regenerated. What do they mean by this?

....

Q2 A garden centre sells two types of compost. 'Paul's Peat' compost is made from peat taken from bogs. 'Sally's Soil' compost is made from decomposed food waste that would otherwise be sent to landfill.

a) i) Explain **one** way that 'Paul's Peat' compost is likely to have a negative effect on biodiversity.

....

....

....

ii) Suggest **one** way that producing 'Sally's Soil' compost could have a positive effect on biodiversity.

....

....

....

b) The company that makes 'Paul's Peat' is worried about the effect they are having on the environment. Other than its effect on biodiversity, why else might the company want to stop using peat in their compost in order to protect the environment? Tick **one** box.

- [ ] **A** Using peat for compost means less peat can be used as fuel.
- [ ] **B** Using peat for compost means more methane is removed from the air.
- [ ] **C** Using peat for compost causes carbon dioxide to be released into the air.

Q3 Some crops are grown in order to produce fuels called biofuels.

a) Tick the correct box to complete the following sentence.

Producing biofuels from crops can reduce biodiversity in the area they are grown because...

- [ ] ... the crops take in carbon dioxide as they grow, helping to reduce global warming.
- [ ] ... forests are often cut down to make way for large fields that contain a single biofuel crop.

b) A farmer is growing a biofuel crop. She allows grasses and wildflowers to grow around the edges of her crop. Suggest why she does this.

....

....

....

Who knew compost could be so controversial...

Our use of land is causing major problems, but it's not all bad news. Lots of people are fighting hard to protect biodiversity and reduce the damage we're causing to the Earth's ecosystems. There's still a lot of work left to do though.

Atoms and Elements

There's no escaping atoms, in chemistry books or in life... They're everywhere.

Warm-Up

All substances are made of atoms. Atoms are tiny, with a radius of just 0.1 nanometers. They're made up of even smaller particles — protons, neutrons and electrons.

Complete the passage below using words from the box:

one	positively	ion	protons	zero
neutrons		element		negatively

Atoms always have a charge of .... An atom which has lost or gained electrons is called an .... A neutral atom has the same number of electrons and .... If an electron is added to a neutral atom, the atom becomes .... charged.

Atoms can be represented by their nuclear symbol. This shows the atomic number, the mass number and the element symbol of the atom. Label the atomic number, mass number and element symbol on the nuclear symbol of fluorine shown below:

.... — $^{19}_{9}F$ — ....

....

I'm in my element!

Q1 **Table 1** contains information about some elements. **Table 1** is incomplete.

Complete **Table 1** using a periodic table.

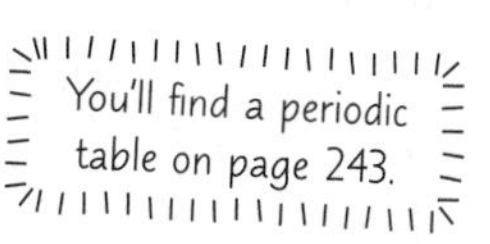

Table 1

Element	Symbol	Mass Number	Number of Protons	Number of Electrons	Number of Neutrons
Sodium	Na	....	11	....	....
Neon	....	....	10	10	10
....	Ca	....	....	20	20

Q2 **Figure 1** shows part of the structure of a neutral atom with a mass number of 11.

a) What is the name for the part of the atom shown in **Figure 1** that is shaded grey?

....

Figure 1

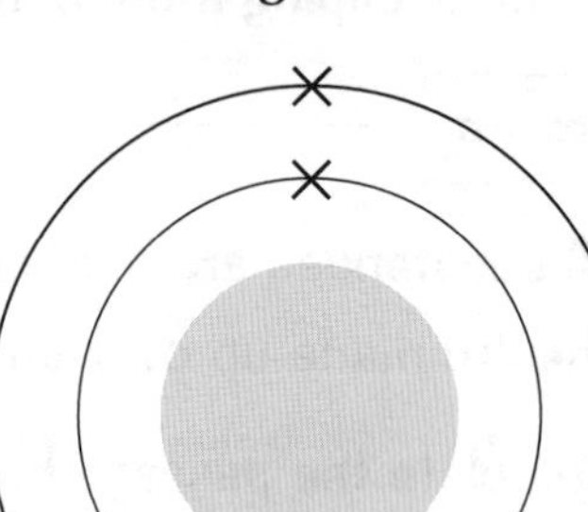

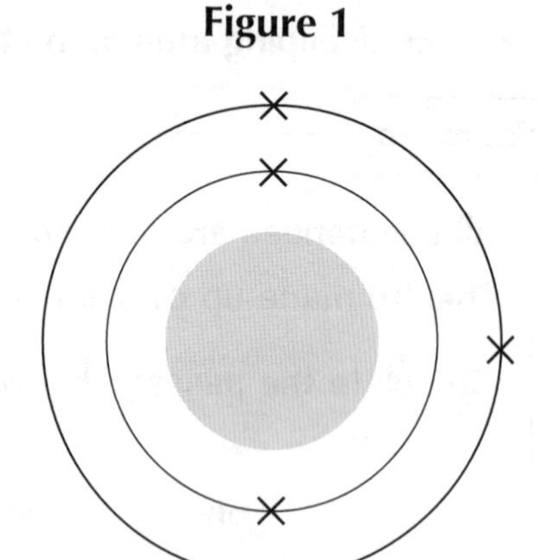

b) **Figure 1** is incomplete.
The atom's electrons are shown on **Figure 1** using ×.
Complete **Figure 1** by adding the correct numbers of protons and neutrons.
Use:

● to show protons,
■ to show neutrons.

c) Use a periodic table to give the name of the element shown in **Figure 1**.

....

Q3 The nuclear symbols for four atoms are shown below.

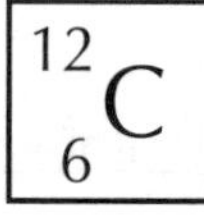

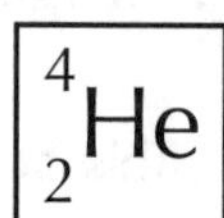

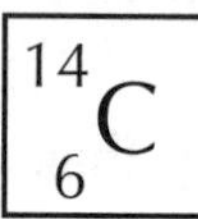

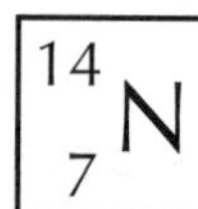

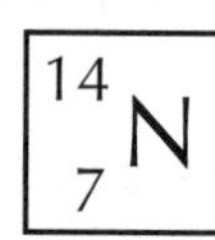

a) Circle the **two** atoms which are isotopes of each other.

b) Explain your answer.

....

....

Q4 Gallium can exist as two stable isotopes: Ga-69 and Ga-71.

60% of gallium atoms are Ga-69 atoms, and the rest are Ga-71 atoms.
Calculate the relative atomic mass (A_r) of gallium to 2 significant figures.
Use the formula below:

$$A_r = \frac{\text{sum of (isotope abundance} \times \text{isotope mass number)}}{\text{sum of abundances of all the isotopes}}$$

Piesotopes

relative atomic mass = ...

Which element has the funniest isotopes? Helium — ³He ⁴He...

Remember, it's the number of protons in an atom that determines which element it is. So atoms of the same element always have the same number of protons, but might have a different number of neutrons — if they do, they're isotopes.

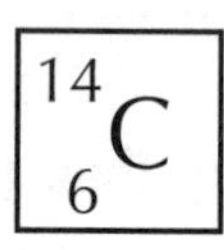

Compounds and Chemical Equations

Elements don't just keep themselves to themselves. Unluckily for you, that means more questions...

Warm-Up

Compounds are formed when two or more elements combine in a chemical reaction. The atoms of the elements are present in set amounts and are held together by chemical bonds. Compounds can only be converted back into elements by chemical reactions.

Look at the following diagrams of substances. Circle the boxes that contain a compound.

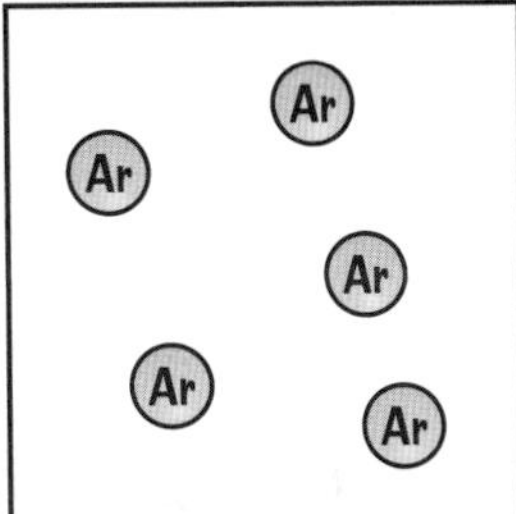

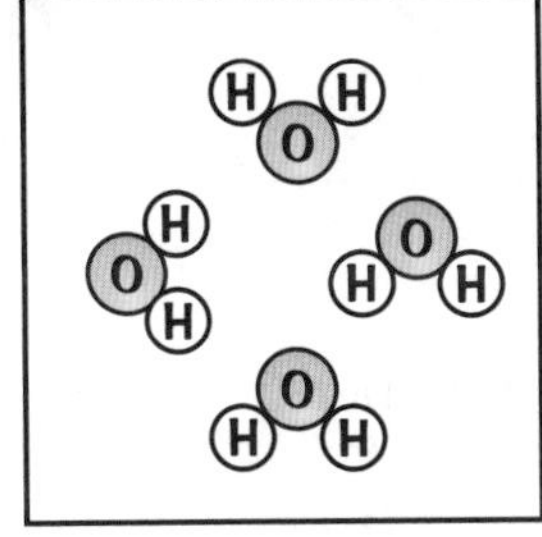

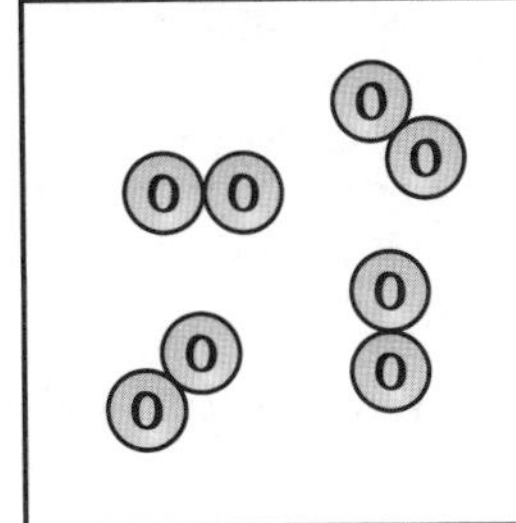

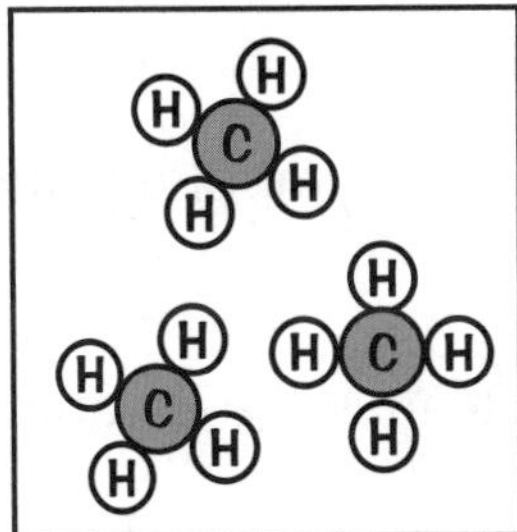

Compounds can be represented by formulas using the symbols of the elements they are made from. Draw lines to match the names of the compounds below with their formulas.

calcium chloride	carbon dioxide	water	sodium hydroxide
CO_2	NaOH	H_2O	$CaCl_2$

Q1 **Table 1** lists four compounds that contain hydrogen.

Complete **Table 1** to show the number of different elements in each compound, and the number of hydrogen atoms in each compound.

Table 1

Compound	Formula	Number of elements	Number of hydrogen atoms
Hydrochloric acid	HCl	....	....
Methane	CH_4	....	....
Calcium hydroxide	$Ca(OH)_2$	....	....
Ethanol	C_2H_5OH	....	....

Q2 Potassium (K) reacts with bromine (Br_2) to give potassium bromide (KBr).

potassium + bromine → potassium bromide

a) i) Name the reactant(s) in this reaction.

....

ii) Name the product(s) in this reaction.

....

b) The picture equation below is unbalanced.
Use it to help you write a balanced symbol equation for the reaction.

You can draw more pictures to help you balance the equation

K + Br Br → K Br

Symbol equation: ....

Q3 Balance these equations by adding in whole numbers.

$Fe_2O_3 + 3CO \rightarrow 2Fe + 3CO_2$

a) $N_2 + 3H_2 \rightarrow$ .... NH_3

b) .... Fe + ... $O_2 \rightarrow 2Fe_2O_3$

c) $4NH_3$ + ... $O_2 \rightarrow 4NO + 6H_2O$

Q4 Lithium (Li) reacts with oxygen (O_2) to form lithium oxide.

a) i) Write a word equation for this reaction.

....

ii) A molecule of lithium oxide contains two atoms of lithium and one atom of oxygen.
Write the formula for lithium oxide.

....

iii) Lewis says: "A molecule of oxygen contains two atoms, so oxygen is a compound."
Explain why Lewis is wrong.

....

....

....

b) Lithium also reacts with water (H_2O) to form lithium hydroxide (LiOH) and hydrogen (H_2). Balance the equation for this reaction.

.... Li + ... $H_2O \rightarrow$ .... LiOH + ... H_2

Sodium and chlorine went on a date — they really bonded...

Balancing equations can be tricky. Keep track of how many of each type of atom you have on each side as you try different numbers in the equation. When both sides match, you'll know you've balanced it.

Mixtures and Chromatography

If you mix two substances that don't react you get... a mixture (creative name, huh?).

Warm-Up

A mixture is made up of two or more substances that are not chemically joined together.

Tick any statements below which are true:

☐ **A** The parts of a mixture can be either elements or compounds.

☐ **B** The chemical properties of a substance are changed if it is part of a mixture.

☐ **C** Mixtures can be separated by carrying out chemical reactions. New substances are produced in the process.

One method for separating mixtures is chromatography.

Q1 John did a paper chromatography experiment to investigate the dyes found in different coloured sweets. The chromatogram he produced is shown in **Figure 1**.

a) Which of the coloured sweets in **Figure 1** definitely contain a mixture of dyes?

....

....

Figure 1

Brown Red Green Orange Blue
Sweet

b) Explain how you can tell.

....

....

....

....

Q2 Magnesium reacts with dilute acid, but copper does not. Sophie has a mixture of small pieces of copper and magnesium.

What will happen if Sophie adds dilute hydrochloric acid to the mixture?

☐ **A** Only the copper will react with the acid.

☐ **B** Both the copper and the magnesium will react with the acid.

☐ **C** Only the magnesium will react with the acid.

☐ **D** Neither the copper nor the magnesium will react with the acid.

Q3 Elena wanted to find out which of five dyes could be present in a particular black ink.

Elena was asked to suggest a method. This is the method she suggested:

1. Take a piece of filter paper. Draw a pencil line near the bottom.
2. Add spots of the dyes to the line at regular intervals.
3. Put the paper into a beaker of water with the line just touching the water.
4. Repeat these steps with a spot of the black ink on a second piece of filter paper.
5. Put this paper into a beaker of ethanol.
6. Place a lid on each beaker. Wait for the solvents to travel to the top of the paper.
7. Compare the positions of the spots created by the black ink with the positions of the spots created by the dyes.

Identify **two** mistakes in this method. For each mistake, suggest how you would alter the method to carry out the experiment correctly.

You can assume Elena takes sensible safety precautions.

Mistake 1: ....

....

Correction: ....

....

....

Mistake 2: ....

....

Correction: ....

....

....

....

Spot the difference.

Why is chromatography so popular? Everyone wants to do it for the 'gram...

There'll be more on chromatography later, but for now make sure you've really nailed how to carry it out. Exactly what you're asked to investigate could vary. So if you're planning your own experiment or describing a method in the exam, make sure that what you've suggested doing will get you the results you need.

More Separation Techniques

PRACTICAL

Didn't think you were getting away with learning just the one, did you?

Warm-Up

Chromatography is one physical process for separating mixtures. There are others you can use depending on what's in the mixture you want to separate. You need to know about simple distillation, fractional distillation, crystallisation and filtration.

Which of the following types of mixture can filtration be used to separate?

☐ A liquids ☐ B soluble solid and liquid ☐ C insoluble solid and liquid

Q1 **Table 1** lists the boiling points of three compounds.

Table 1

Name	Boiling point (°C)
diethyl ether	35
THF	66
ethyl ethanoate	77

a) Which techniques could be used to separate a mixture of diethyl ether and ethyl ethanoate? Tick **two** boxes.

☐ **A** filtration
☐ **C** crystallisation
☐ **B** simple distillation
☐ **D** fractional distillation

b) Which technique could be used to separate a mixture of ethyl ethanoate and THF? Tick **one** box.

☐ **A** filtration
☐ **C** crystallisation
☐ **B** simple distillation
☐ **D** fractional distillation

c) Explain why two different techniques can be used to separate diethyl ether and ethyl ethanoate, but only one technique can be used to separate ethyl ethanoate and THF.

....

....

....

....

....

Q2 Mei is using crystallisation to obtain a sample of solid potassium nitrate.

She begins by gently heating potassium nitrate solution in an evaporating dish until crystals start to form. Describe how she should complete the process.

....

....

....

....

Q3 The boiling point of methanol is 65 °C. The boiling point of propanol is 97 °C.

Jo is using the apparatus shown in **Figure 1** to separate a mixture of methanol and propanol.

Figure 1

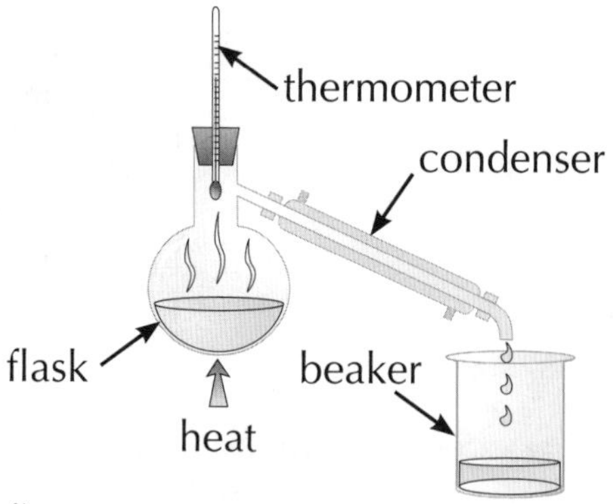

thermometer
condenser
flask
heat
beaker

a) i) At the start of the experiment, the flask contains the mixture of methanol and propanol. What will the contents of the flask and the beaker be at the end of the experiment?

Contents of the flask: ....

Contents of the beaker: ....

ii) Give a reason for your answer to i).

....

....

....

Unexpected flask contents...

b) Jo then continues to heat the flask until the reading on the thermometer is 104 °C. The beaker contains liquid, but the flask does not. Explain why Jo has not successfully separated the methanol and propanol.

....

....

....

Just don't do it by text, whatever you do...

Phew, I feel like separating myself from this book after all of that. If you found that you kept getting the techniques mixed up in these questions, read over your notes again until it all becomes crystal clear and have another go.

The History of the Atom

I know it says history, but I promise there's no essay-writing involved here.

Warm-Up

Over the past two hundred years, changes in the <u>theory of atomic structure</u> have come about as a result of <u>new experimental evidence</u>.

For example, the discovery of the neutron in 1932 showed that the <u>nucleus</u> was made up of <u>two</u> types of <u>smaller particles</u> — <u>protons</u> and <u>neutrons</u>.

The Bore model...

Who discovered the neutron?

....

Q1 **Table 1** shows two past discoveries made about the structure of the atom.

Table 1

Discovery	Model of the atom that was used until the discovery	Model of the atom that was developed using the discovery
The discovery of the electron.	....	....
When alpha particles are fired at a thin sheet of gold, some are deflected more than expected. Some are even deflected backwards.	....	....

Complete **Table 1** using the boxes below to show which model of the atom existed before each discovery and which model of the atom was developed after each discovery.
You may not need to use every box.

The nuclear model	The plum pudding model	Atoms are solid spheres	The Bohr model

Q2 Dylan and Zara draw diagrams to show the different models of the atom.

a) **Figure 1** shows Dylan's labelled diagram of the plum pudding model. Dylan has made two mistakes in his diagram. Identify these mistakes and describe how Dylan should correct them.

Figure 1

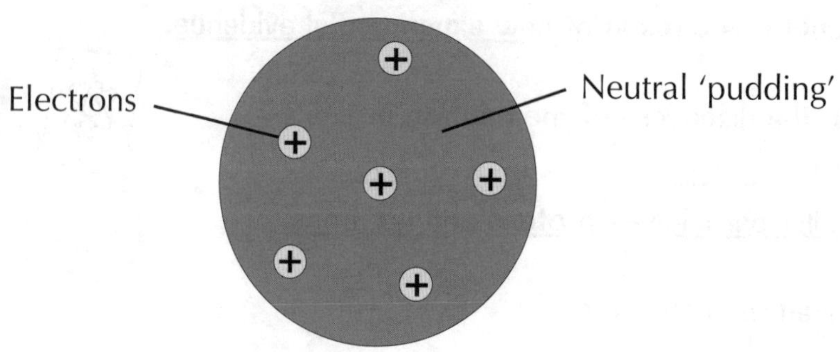

Mistake 1: ....

....

Correction: ....

....

Mistake 2: ....

....

Correction: ....

....

b) Zara is drawing the Bohr model of the atom. **Figure 2** shows her incomplete diagram. Complete **Figure 2** so that it shows the Bohr model.

Figure 2

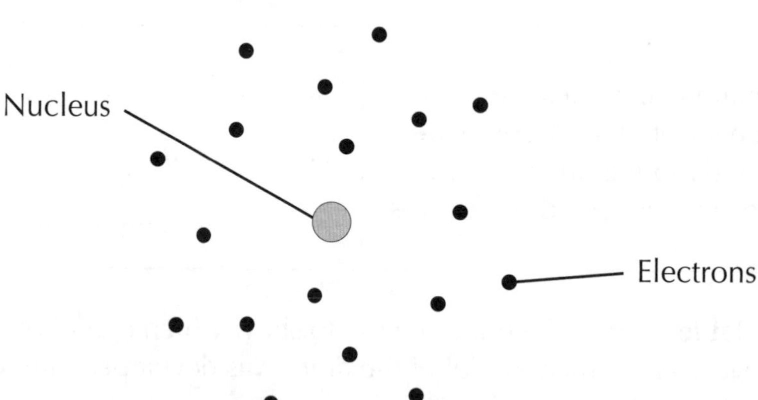

All this talk of pudding is making me hungry...

It's really important you don't just learn <u>how</u> the model of the atom changed over time — you also need to know <u>why</u>. It's a classic example of the scientific method in action — a theory can only last as long as it can explain all available evidence.

Electronic Structure

Ok, so electrons are found in shells, but how are they arranged?

Warm-Up

In an atom, <u>electrons</u> always move around the nucleus in fixed <u>shells</u>.
These shells are sometimes called <u>energy levels</u>.

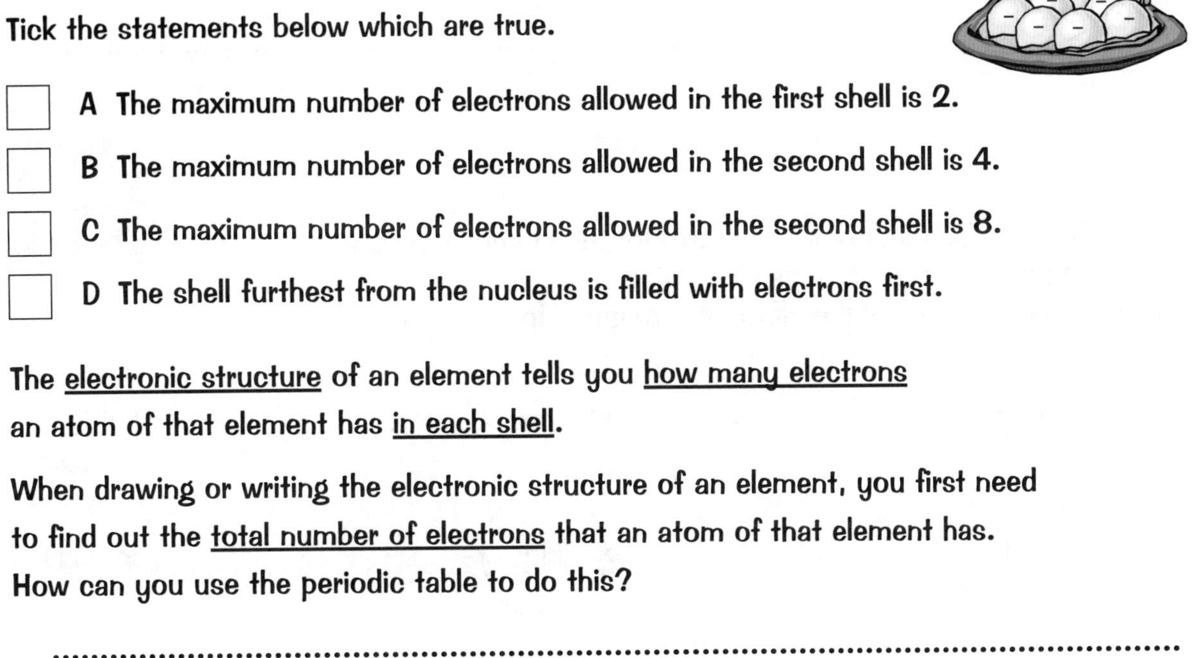

There are a set maximum number of electrons allowed in each shell.
Tick the statements below which are true.

☐ **A** The maximum number of electrons allowed in the first shell is 2.

☐ **B** The maximum number of electrons allowed in the second shell is 4.

☐ **C** The maximum number of electrons allowed in the second shell is 8.

☐ **D** The shell furthest from the nucleus is filled with electrons first.

The <u>electronic structure</u> of an element tells you <u>how many electrons</u> an atom of that element has <u>in each shell</u>.

When drawing or writing the electronic structure of an element, you first need to find out the <u>total number of electrons</u> that an atom of that element has.
How can you use the periodic table to do this?

....

....

....

Q1 **Figure 1** shows the electronic structure of potassium.
Figure 2 shows part of the electronic structure of silicon. **Figure 2** is not complete.

Figure 1

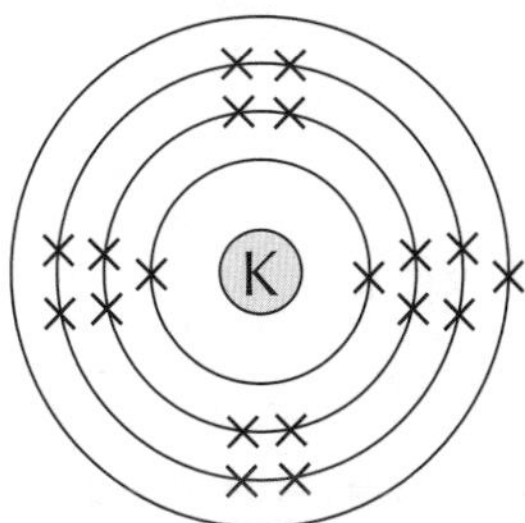

Figure 2

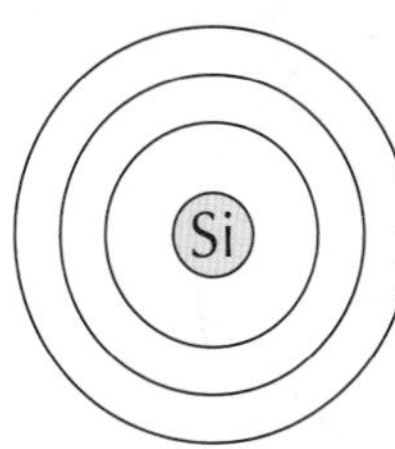

a) Use **Figure 1** to write the electronic structure of potassium in number form.

....

b) Draw an arrow on **Figure 1** pointing to the shell that gets filled with electrons first.

c) The electronic structure of silicon is 2, 8, 4.
Complete **Figure 2** to show the electronic structure of silicon.

Q2 Neon has 10 electrons. **Figure 3** incorrectly shows the electronic structure of neon.

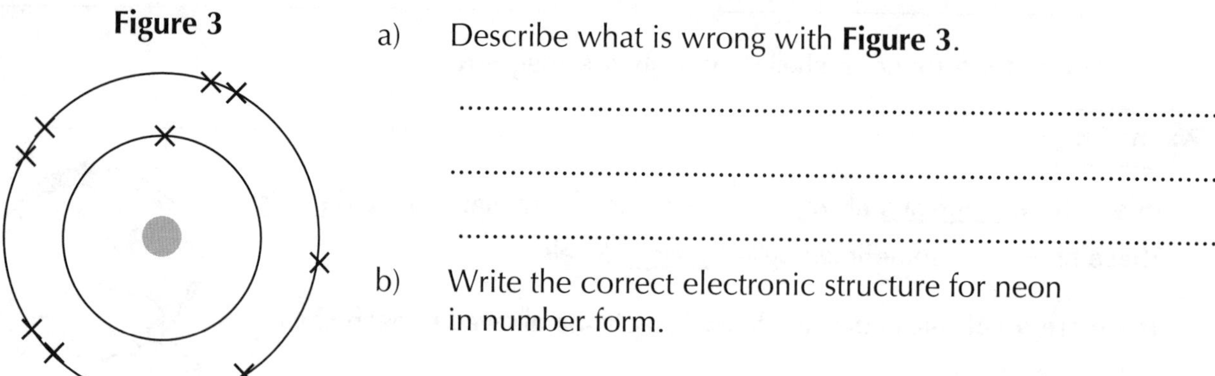

Figure 3

a) Describe what is wrong with **Figure 3**.

....

....

....

b) Write the correct electronic structure for neon in number form.

....

An electronic structure

Q3 Complete the full electronic structures for the elements below.
The first three have been done for you.
The atomic numbers of the elements are given in brackets.

Hydrogen (1)

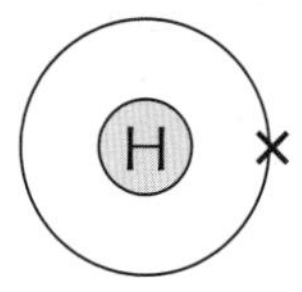

Helium (2)

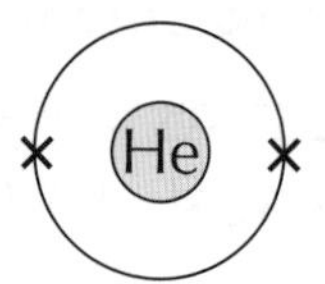

Lithium (3)

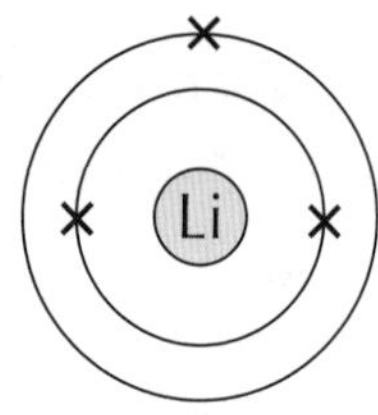

a) Boron (5)

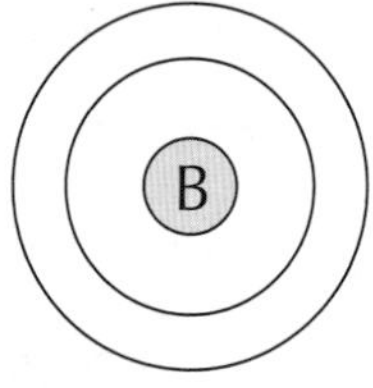

b) Oxygen (8)

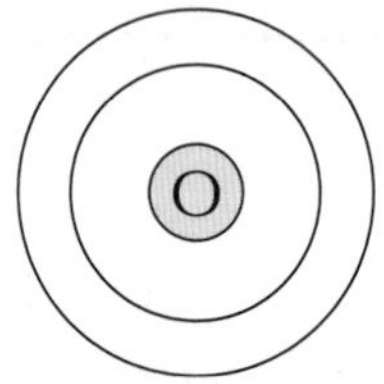

c) Fluorine (9)

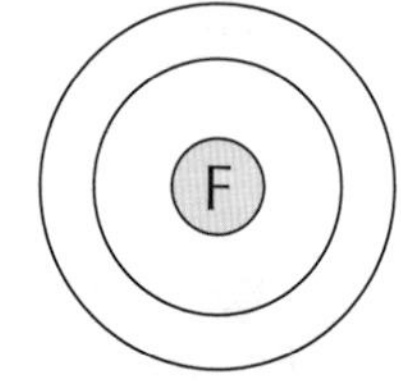

d) Sodium (11)

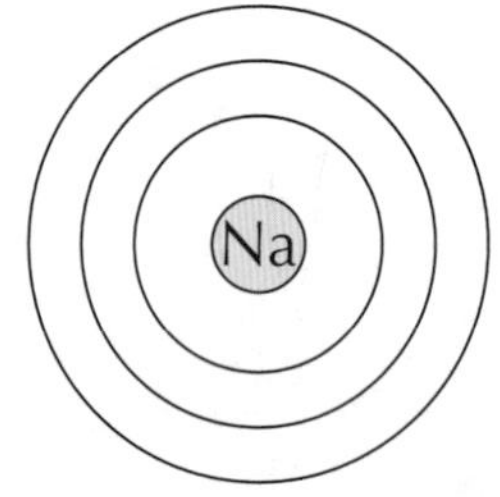

e) Aluminium (13)

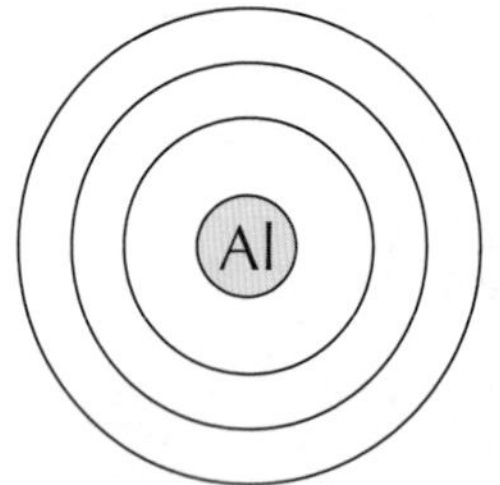

f) Argon (18)

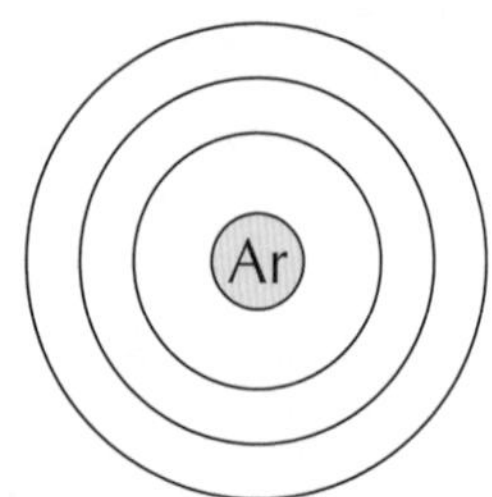

The 7 am alarm — that's when I'm at my lowest energy level...

Ok, so all of these diagrams might look a little scary, but when you break it down it's just about following a set of simple rules. As long as you remember to fill your shells one at a time and not to overfill them, life's a beach.

The Periodic Table

It's more than just a pretty poster on your classroom wall...

Warm-Up

The periodic table was developed by chemists to help them understand patterns in the properties of elements.

Use the words in the box below to complete the passage about the periodic table.

groups	reactivity	electrons	atomic number	
different	atomic mass	periods	similar	protons

Early periodic tables were produced by placing the elements in order of .... In the modern periodic table the elements are arranged in order of .... The columns in the table are called .... These columns contain elements with .... chemical properties. Elements in the same column all have the same number of .... in their outer shell.

In early periodic tables, some elements were placed in the wrong groups. Which of the following things did Dmitri Mendeleev do to make sure elements were in the correct group? Tick two boxes.

- [ ] **A** He used the atomic numbers of the elements to put them in order.
- [ ] **B** He switched the order of some elements to make sure they were in groups with elements with similar properties.
- [ ] **C** He left gaps in the table to make sure elements with similar properties stayed in the same groups.
- [ ] **D** He removed the elements that were in the wrong groups.

Q1 Select from the elements below to answer the following questions.

iodine nickel phosphorus sodium radon krypton calcium

a) Which **two** elements are in the same group?

.... and ....

b) Name **two** elements which are in Period 3.

.... and ....

c) Name an element in Group 1.

....

A period drama

Q2 **Table 1** shows some information about selected elements from the periodic table.

Table 1

Name	Group number	Period number	Electronic structure
....	4	....	2, 4
....	....	3	2, 8, 4
Nitrogen	5	....	2, 5
....	6	....	2, 8, 6
Beryllium	....	2	2, 2

a) Complete **Table 1**.

b) i) How are the group numbers of the elements related to their electronic structures?

....

....

ii) How are the period numbers of the elements related to their electronic structures?

....

....

Q3 Beth and Aaliyah are investigating the reactions of sodium, potassium and magnesium with water.

They start by adding a piece of sodium metal to water.
The sodium reacts and a gas and a colourless solution are formed.
Beth and Aaliyah predict which of the other metals will react most similarly to the sodium.

Magnesium will react most similarly, as it's in the same period as sodium and its atomic number is only one higher.

Beth

Potassium will react most similarly, because it's in the same group as sodium.

Aaliyah

Who is correct? Explain why she is correct.

....

....

....

Why couldn't hydrogen enter a pop band competition? It's not in a group...

The periodic table is a really handy way of displaying a load of information about the elements. You'll get given a copy in the exam, but it's worth getting familiar with it now so you can get the information you need when you need it.

Metals and Non-Metals

Doom, death, thrash, folk, Viking... I know my metals.

Warm-Up

Non-metals are found at the far right and top of the periodic table.
They tend to either share or gain electrons to get a full outer shell.
Metals are elements found to the left and towards the bottom of the periodic table.

Circle the correct words to complete the following passage about metals.

Metals to the left of the periodic table have **many / few** electrons to remove in order to be left with a full outer shell. Metals towards the bottom of the periodic table have outer electrons which are **close to / far away from** the nucleus and so feel a **stronger / weaker** attraction to it. This means that **not much / a lot of** energy is needed to remove electrons from these metals.

Q1 Read the descriptions of elements **X**, **Y** and **Z** and answer the questions that follow.

Element **X** is soft and brittle. It is a poor conductor of heat.
It has a melting point of 115 °C and a boiling point of 445 °C.

Element **Y** is soft and shiny.
It conducts heat well, and can be easily bent into different shapes.
It has a melting point of 660 °C and a boiling point of 2519 °C.

Element **Z** is hard and shiny.
It conducts heat well, and can be easily bent into different shapes.
It has a melting point of 1728 °C and a boiling point of 3003 °C.

a) i) Which of the elements, **X**, **Y** and **Z**, is most likely to be a non-metal?

....

ii) Give **two** reasons for your choice.

1. ....

2. ....

....

b) Which **two** of the three elements, **X**, **Y** and **Z**, will form positive ions?

....

Q2 **Table 1** shows the properties of four elements found in the periodic table.

Table 1

Element	Melting point (°C)	Density (g/cm³)	Electrical conductivity
A	1084	8.90	Excellent
B	1064	19.3	Excellent
C	115	2.07	Very poor
D	1536	7.87	Very good

a) Which **three** of the elements in **Table 1** are most likely to be metals?

....

b) Explain why the other element is least likely to be a metal.

....

....

....

Q3 Mohammed carried out an experiment using the apparatus shown in **Figure 1**.
He had four identically-sized rods (**A**, **B**, **C** and **D**), each made from a different material.
He heated each rod at one end, and connected a temperature sensor to the other end.

For each rod, Mohammed measured the time taken for the temperature recorded by the sensor to increase by a certain amount.
Figure 2 shows the order of the speed in which the rods heated up.

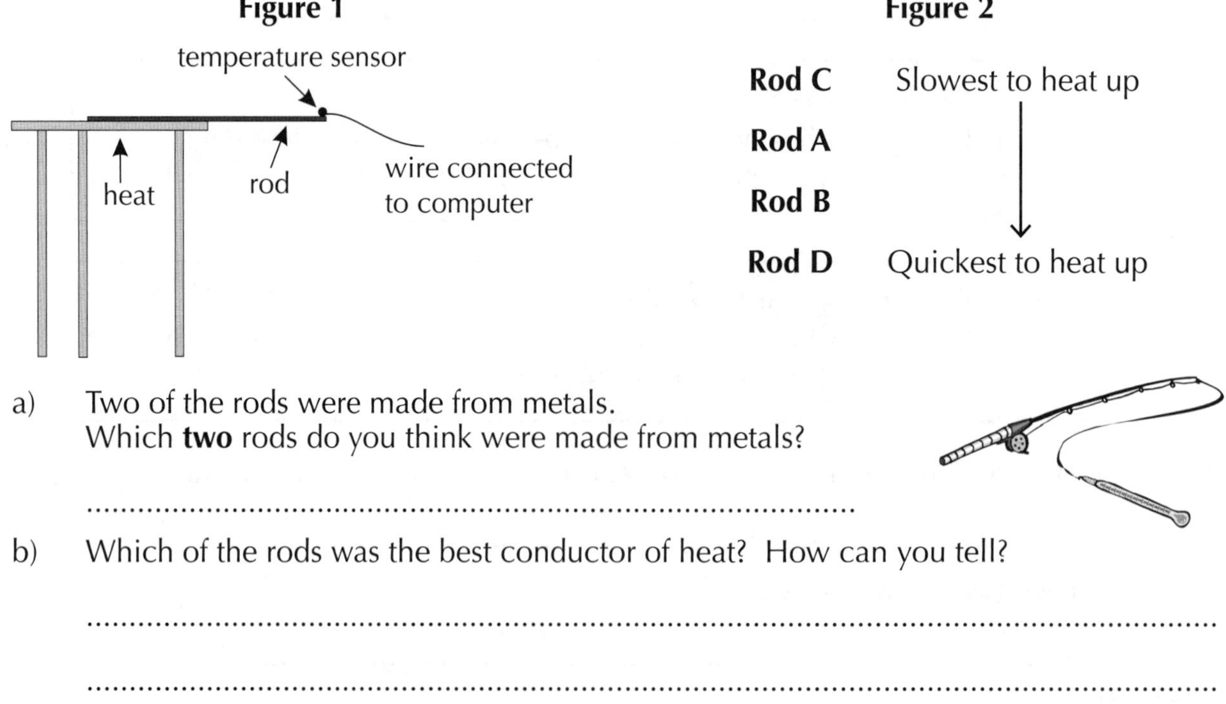

a) Two of the rods were made from metals.
Which **two** rods do you think were made from metals?

....

b) Which of the rods was the best conductor of heat? How can you tell?

....

....

These questions will really test your metal...

If you're finding it hard to remember the properties of metals, try linking them to some of the metal objects you use in everyday life. For example, metals are used to make saucepans because they have high melting points (so won't melt all over your hob) and are good conductors of heat (so your food actually gets cooked).

Group 1 Elements

Time to look at one of the groups from the periodic table in more detail...

Warm-Up

The elements in Group 1 are known as the alkali metals.
They each have one electron in their outer shell, which makes them very reactive.

Circle the correct words to complete the passage below about alkali metal reactions.

It doesn't take much energy for alkali metals to lose their outer shell electron. This means they can easily form **1+ / 1–** ions. Because they form ions so easily, alkali metals easily form **covalent / ionic** compounds. Alkali metals react with water to produce **hydrogen / oxygen** gas and a **hydroxide / chloride** solution.

Q1 The elements of Group 1 show trends in their properties.

a) Choose an element from the box below to answer each of the following questions. Use the periodic table to help you.

rubidium	sodium	potassium	lithium	francium	caesium

i) Which element has the highest relative atomic mass?

....

ii) Which element is the least reactive element?

....

b) The melting and boiling points of the Group 1 elements decrease down the group.

i) The melting point of sodium is 98 °C. The melting point of rubidium is 39 °C. Which of the following could be the melting point of potassium? Tick **one** box.

☐ **A** 26 °C ☐ **B** 63 °C ☐ **C** 124 °C

ii) The boiling point of lithium is 1342 °C. The boiling point of rubidium is 688 °C. Which of the following could be the boiling point of sodium? Tick **one** box.

☐ **A** 661 °C ☐ **B** 1507 °C ☐ **C** 883 °C

Q2 A piece of lithium is heated in chlorine gas.

a) Write a word equation for the reaction that takes place.

....

b) A piece of potassium is heated in chlorine gas.
Would the reaction be more or less vigorous than the reaction with lithium?

....

Q3 Archibald dropped samples of three different alkali metals, **A**, **B** and **C**, into bowls of water. Each sample has the same mass and surface area. In each case, the metal reacted with the water and disappeared.

a) The time taken for each metal to disappear is shown in **Table 1**.

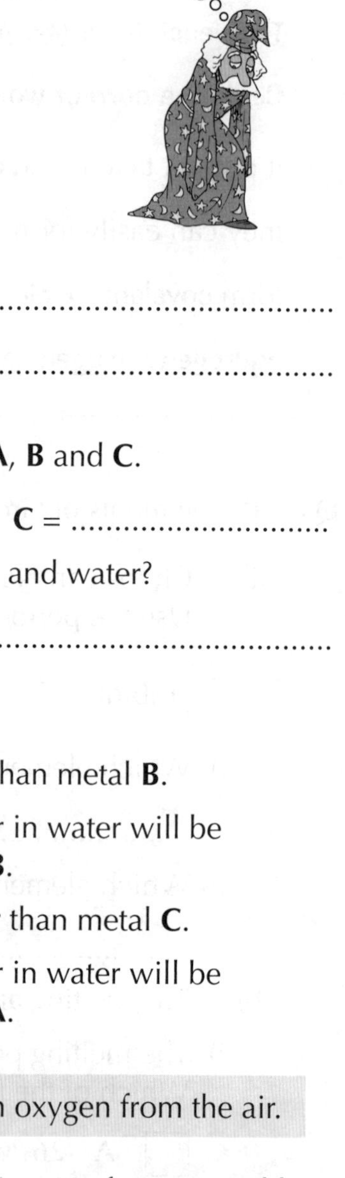

Table 1

Metal	Time taken to disappear (s)
A	27
B	8
C	42

i) Which of the metals in **Table 1** is the most reactive? How can you tell?

....

....

ii) The three metals used were lithium, sodium and potassium. Use the results in **Table 1** to work out the identity of metals **A**, **B** and **C**.

A = ... **B** = ... **C** = ...

b) i) What products were formed in the reaction between sodium and water?

....

ii) Rubidium also reacts with water. Which of the following statements is correct? Tick **one** box.

- [ ] **A** Rubidium will take less time to disappear in water than metal **B**.
- [ ] **B** The amount of time taken for rubidium to disappear in water will be shorter than for metal **A**, but longer than for metal **B**.
- [ ] **C** Rubidium will take more time to disappear in water than metal **C**.
- [ ] **D** The amount of time taken for rubidium to disappear in water will be shorter than for metal **C**, but longer than for metal **A**.

Q4 Alkali metals should be stored under oil to stop them reacting with oxygen from the air.

A scientist finishes working with a sodium sample and puts it in a jar. He forgets to add oil to the jar. When he next wants to use the sample, he notices that the surface has changed from a shiny silver to a dull grey. Explain what has happened to the sodium.

....

....

....

Want to hear a joke about potassium? K...

Luckily for you, the reactions of Group 1 elements with water and with chlorine follow the same patterns. So if you know the word and balanced symbol equations for one Group 1 element, you actually know them all. All you need to do is swap out the name or symbol of that Group 1 element for the one you need. Simple.

Group 7 Elements

Skip over Groups 2-6 — Group 7's the next one you need to know about.

Warm-Up

The elements of Group 7 are known as the halogens. The halogens are non-metals that exist as molecules which are pairs of atoms. Just like in Group 1, there are trends in the properties of the elements as you move down Group 7.

Write the halogens from the box below in order of increasing relative molecular mass.

Iodine	Fluorine	Bromine	Astatine	Chlorine

....

Lowest relative molecular mass — Highest relative molecular mass

Draw lines to match the first three halogens to their melting points.

Chlorine	Bromine	Fluorine
–7 °C	–220 °C	–101 °C

Q1 Sodium metal was reacted with bromine vapour. **Figure 1** shows the apparatus used. White crystals of a new solid were formed during the reaction.

Figure 1

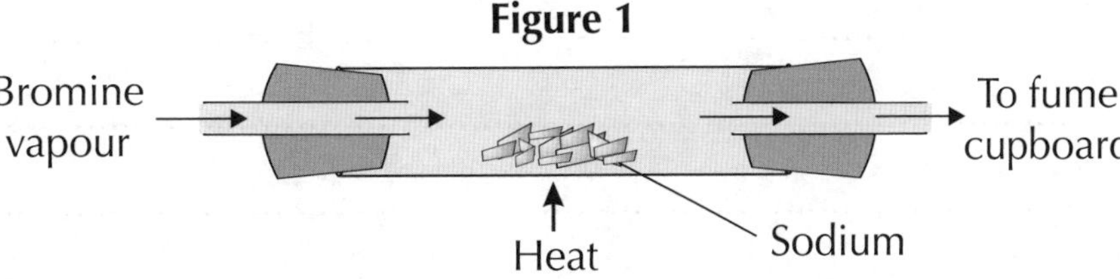

a) What type of bonding is present in the white crystals that were formed?

....

b) The reaction in **Figure 1** was repeated under the same conditions, first using iodine instead of bromine, and then using chlorine. For each of these reactions, state whether it would be faster or slower than the reaction shown in **Figure 1**.

i) The reaction between sodium and iodine vapour.

....

ii) The reaction between sodium and chlorine gas

....

Q2 Fluorine, F_2, reacts with hydrogen, H_2, to form hydrogen fluoride, HF.

a) What type of bond is present in hydrogen fluoride?

....

b) Fluorine reacts explosively with hydrogen at low temperatures. Iodine reacts slowly and incompletely with hydrogen when heated strongly. The passage below explains this difference in reactivity. Use words from the box to complete the passage.

inner	more	negative	above	less
gain	below	outer	lose	beside

Iodine is .... fluorine in Group 7, and so its

.... electron shell is further from its nucleus.

This makes it harder for iodine to .... an electron,

and so it is .... reactive than fluorine.

This makes iodine react more slowly and incompletely.

Oh hallo Jen — I'll call you back...

Q3 Equal volumes of bromine water were added to two test tubes. Each test tube contained a different potassium halide solution. The observations from the test tubes are shown in **Table 1**.

Table 1

Solution	Observations
potassium chloride	no reaction
potassium iodide	reaction took place

a) Explain these observations.

....

....

....

....

b) Complete the word equation for the reaction in the iodide solution.

bromine + potassium iodide → .... + ...

c) Would there be a reaction between bromine water and potassium fluoride?

....

I'm just an average kind of guy — I'm the bro-mean...

You won't need to remember the individual melting or boiling points of any of the elements, but you could be asked to predict some of them using given information. So make sure you understand how they change as you move down the group.

Group 0 Elements

What comes after Group 7? Group 0 of course...

Warm-Up

The Group 0 elements are also known as the noble gases. They are found on the far right of the periodic table. They're all unreactive gases at room temperature.

Tick the statements below which are true:

☐ A The noble gases are non-metals.

☐ B The noble gases exist as molecules made of pairs of atoms.

☐ C The noble gases have full outer electron shells.

☐ D The noble gases easily form both positive and negative ions.

Q1 **Figure 1** shows the outer shell electron arrangements of five atoms, **A-E**.

Figure 1

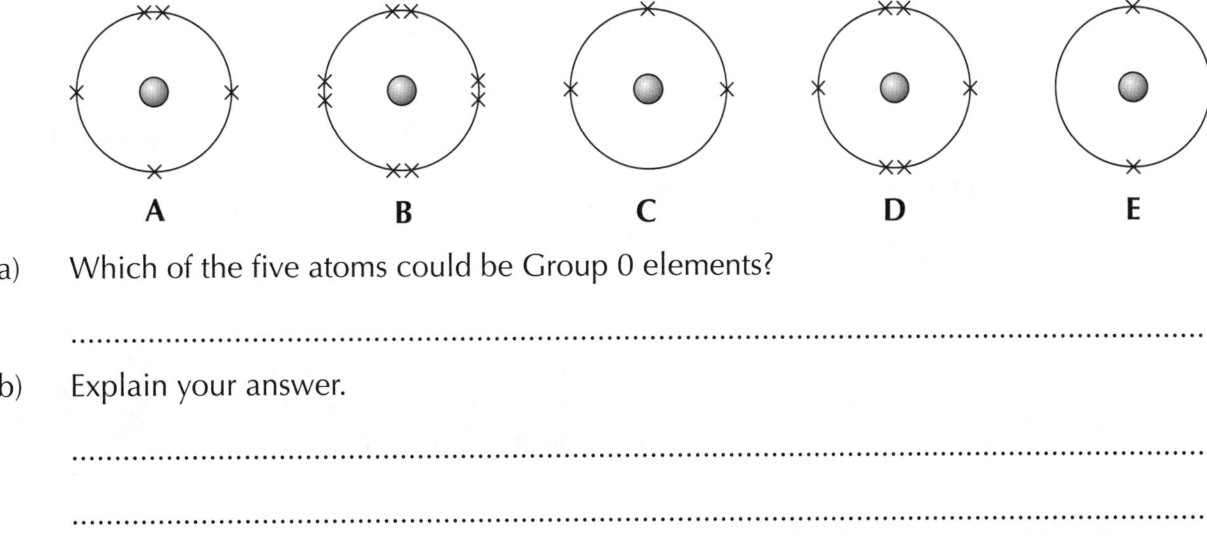

a) Which of the five atoms could be Group 0 elements?

....

b) Explain your answer.

....

....

Q2 Mariya tries to burn a sample of neon gas in air.

What happens? Tick **one** box.

☐ **A** The neon burns with a bright white flame.

☐ **B** The neon reacts with nitrogen from the air.

☐ **C** The neon atoms lose electrons and become 1+ ions.

☐ **D** Nothing.

Q3 There are trends in the properties of the Group 0 elements.

a) i) Complete **Table 1** to show the relative atomic masses and the boiling points of the Group 0 elements. Use the numbers in the boxes on the left.

4 | −246 | −186 | 40

Table 1

Element	Relative atomic mass	Boiling point (°C)
Helium	....	−269
Neon	20	....
Argon	....	....

ii) The melting points of the elements increase moving down Group 0. Argon is a solid at −200 °C. Predict the state of krypton at −200 °C. Explain your prediction.

Prediction: ....

Explanation: ....

....

....

b) The densities of the Group 0 elements increase as you go down Group 0. **Table 2** shows the densities of helium and argon at 20 °C.

Table 2

Element	Density (g/cm^3)
Helium	0.0002
Argon	0.0018

Predict the density of neon at 20 °C.

.... g/cm^3

Q4 Some light bulbs contain a thin metal filament (wire). If these bulbs were filled with air, oxygen would react with the filament and cause it to burn away. To avoid this, the light bulbs are filled with argon.

Explain why argon is suitable for this use.
You should talk about the outer shell electrons of an argon atom in your answer.

....

....

....

....

Trying to think of a joke for this page, but my wit and creativity argon...

So, surprise surprise, the Group 0 elements show trends in their properties, just like the elements in Groups 1 and 7. Thankfully, they don't follow any of the trends that I did back in the 2000s, and they are much easier to predict.

Ions and Ionic Bonding

Ions sound pretty space age. They're even more exciting than that, believe me.

Warm-Up

Any room in that shell, noble snail?

Sorry, it's full.

Ions are particles which have a charge.
They form when atoms gain or lose electrons. They do this so they can have a full outer shell of electrons — like a noble gas.

A metal atom can transfer the electrons it loses to a non-metal atom. This forms a positive metal ion and a negative non-metal ion. The two oppositely charged ions are attracted to each other by electrostatic forces. This is called an ionic bond.

Sort the elements on the left into the correct column of the table on the right.

lithium magnesium

caesium sulfur

bromine fluorine

Forms a positive ion	Forms a negative ion

Q1 Which of the following diagrams shows an oxide ion (O^{2-}) forming from an oxygen atom?

☐ **A** ☐ **B**

☐ **C** ☐ **D**

Q2 Different atoms need to gain or lose different numbers of electrons to get a full outer shell.

a) How many electrons do the following elements need to lose to get a full outer shell? Write your answers in the boxes.

lithium ☐ calcium ☐ potassium ☐

b) How many electrons do the following elements need to gain to get a full outer shell? Write your answers in the boxes.

sulfur ☐ chlorine ☐ fluorine ☐

Q3 Rhodium is an element which can form 3+ ions.
Is rhodium a metal or a non–metal? Explain your answer.

....

....

Q4 Potassium selenide, K_2Se, is an ionic compound.
Potassium (K) is in Group 1 and selenium (Se) is in Group 6.

Draw a dot and cross diagram to show the bonding in potassium selenide.
You only need to draw the outer shell electrons. Include both charges.

Q5 Dervla sees a dot and cross diagram of caesium chloride (CsCl) and says the following:

Ions in caesium chloride are arranged in the same way as ions in sodium chloride.

Why is Dervla wrong to draw this conclusion from this dot and cross diagram?

....

....

Q6 Astatine (At) is in Group 7 of the periodic table.

a) Draw a dot and cross diagram to show the electronic structure of an astatine ion.
You only need to draw the outer shell electrons. Include the charge of the ion.

b) Which atom in Group 0 has the same electronic structure as an astatine ion?

....

I'll happily ion your shirts — there's a charge though...

These ideas can seem a bit confusing at first, but with practice, they definitely get easier. Make sure you get loads of practice at drawing dot and cross diagrams and that you can describe how ionic bonds are formed.

Ionic Compounds

Ooh bionic compounds, this should be good. Oh, wait — this is ionic compounds. My mistake.

Warm-Up

In an ionic compound, oppositely charged ions are held together closely by very strong electrostatic forces of attraction. The strong forces of attraction between ions give ionic compounds similar properties.

Tick the box next to each statement that is true.

☐ **A** Ionic compounds conduct electricity in all states.

☐ **B** Ionic compounds only conduct electricity when molten.

☐ **C** Ionic compounds have high melting points.

☐ **D** Ionic compounds don't melt.

Life's great.

Rubbish — everything is terrible.

Q1 Diagrams can be used to represent the structures of chemical substances.

a) Lithium chloride has a similar structure to sodium chloride. Which of the following diagrams could be used to show how ions bond in solid lithium chloride?

☐ **A** ☐ **B** ☐ **C** ☐ **D**

b) What type of structure does solid lithium chloride have?

....

c) Jade draws a ball and stick diagram to show the structure of lithium chloride. Tick the correct box next to each statement below to show whether it is **true** or **false**.

	True	False
i) The ball and stick diagram would show how big the lithium and chloride ions are compared to each other.	☐	☐
ii) The ball and stick diagram will be accurate because it will show the gaps between lithium and chloride ions.	☐	☐
iii) The ball and stick diagram will show the ordered pattern of lithium and chloride ions.	☐	☐

Q2 Some elements and the ions they form are shown below.

beryllium, Be^{2+}	potassium, K^{+}	iodine, I^{-}	sulfur, S^{2-}

Write the formulas of four ionic compounds which can be made using just these elements.

1. .... 2. ....

3. .... 4. ....

Q3 Potassium chloride is a salt that can be found in the sea. **Table 1** shows some information about how potassium chloride conducts electricity.

a) Circle the correct options in **Table 1** to show the properties of potassium chloride.

Table 1

	When solid	When dissolved in water
Conducts electricity?	Yes / No	Yes / No

b) Explain your answers to part a).

....

....

....

Q4 **Figure 1** shows the structure of iron(II) oxide. In forming the compound, the iron atoms lost two electrons each.

Figure 1

Oxide ions are formed from oxygen atoms.
○ Oxide ions
● Iron ions

a) What are the charges of the ions in iron(II) oxide?

Iron ions: ....

Oxide ions: ....

b) What is the empirical formula of iron(II) oxide?

....

c) Explain whether or not you would expect iron(II) oxide to have a low melting point.

....

....

....

One giant ionic salad — no tomato but plenty of lattice please...

If you're told a compound is ionic, you can usually predict its physical properties. So whether you have sodium chloride, rubidium bromide or even barium iodide, you know it's probably going to have a high boiling point. How very handy.

Covalent Bonding

Share your chocolate with someone new. If they share some with you too, you've formed a bond.

Warm-Up

Covalent bonds are formed when atoms share electrons in bonds.

A covalent bond forms between...

...a metal atom and a non-metal atom. ☐ ... two non-metal atoms. ☐

...two metal atoms. ☐ ...an ion and a non-metal. ☐

The positively charged nuclei are attracted to the shared pair of electrons through electrostatic forces. Covalent bonds are very strong.

Each atom will form enough covalent bonds to...

...empty its outer shell of electrons. ☐ ...fill its outer shell of electrons. ☐

Dot and cross diagrams can be used to show the covalent bonds in a molecule.

Cross Dot

Q1 Silicon has the electronic structure 2, 8, 4.

a) What is the maximum number of covalent bonds that silicon will form in a simple covalent molecule? Tick **one**.

☐ **A** none ☐ **B** two ☐ **C** four ☐ **D** eight

b) Explain your answer to part a).

....

....

Q2 **Figure 1** shows a molecule of hydrazine.

Figure 1

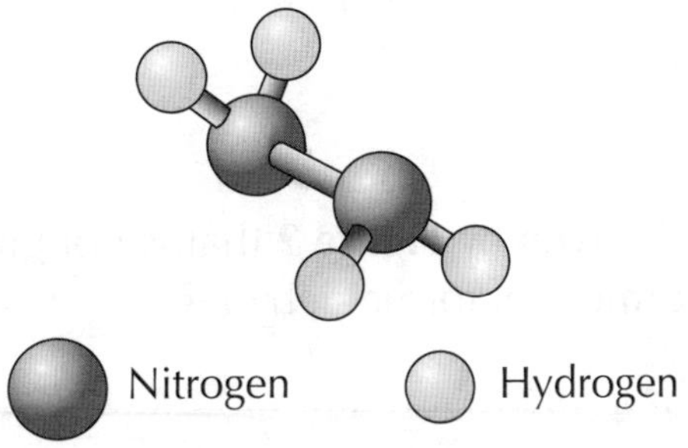

Nitrogen Hydrogen

a) Write down the molecular formula of hydrazine.

....

b) Give **one** advantage of using the model shown in **Figure 1** to represent the structure of hydrazine over using a dot and cross diagram.

....

Q3 Which of the following diagrams does not represent the structure of bromomethane, CH_3Br?

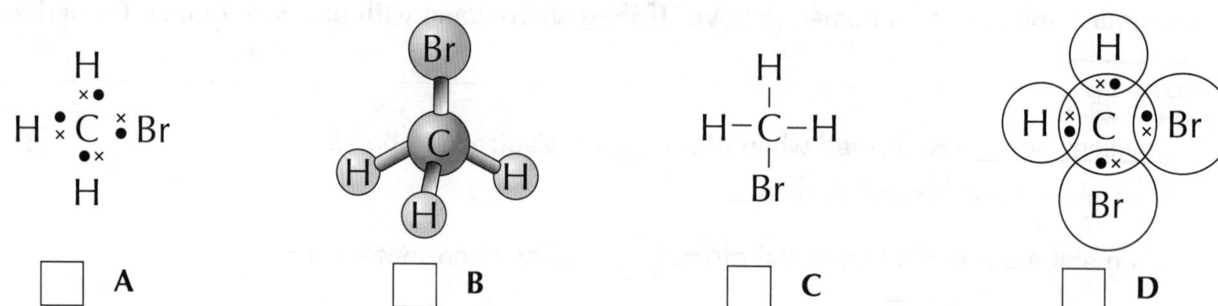

Q4 The bonding in formaldehyde is represented in **Figure 2**.

Figure 2

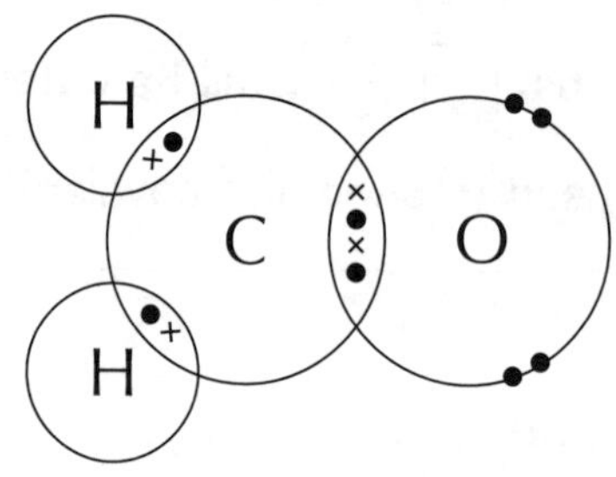

a) What is the molecular formula of formaldehyde?

....

b) How many covalent bonds has carbon formed in formaldehyde?

....

c) Draw the displayed formula of formaldehyde in the space below.

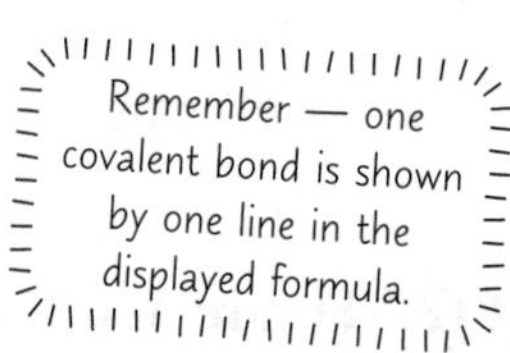

d) What information is given by **Figure 2** that is not given by the displayed formula of formaldehyde?

....

....

Terrible covalent bonding jokes — best not to be shared....

Covalent bonds are the bread and butter of a lot of chemistry, so you've got to make sure you understand them well. Make sure you can explain how they form. Practise drawing dot and cross diagrams too — they're pretty important.

Simple Molecular Substances

Simple by name, simple by nature. And with practice, answering these questions will be simple too.

Warm-Up

Simple molecular substances are made up of molecules containing a small number of atoms. These atoms are joined by covalent bonds.
The forces of attraction between simple molecules are generally very weak, but their strength depends on the size of the molecule.
Simple molecular compounds can be represented using dot and cross diagrams.

Complete the dot and cross diagrams of the simple molecular compounds shown below.

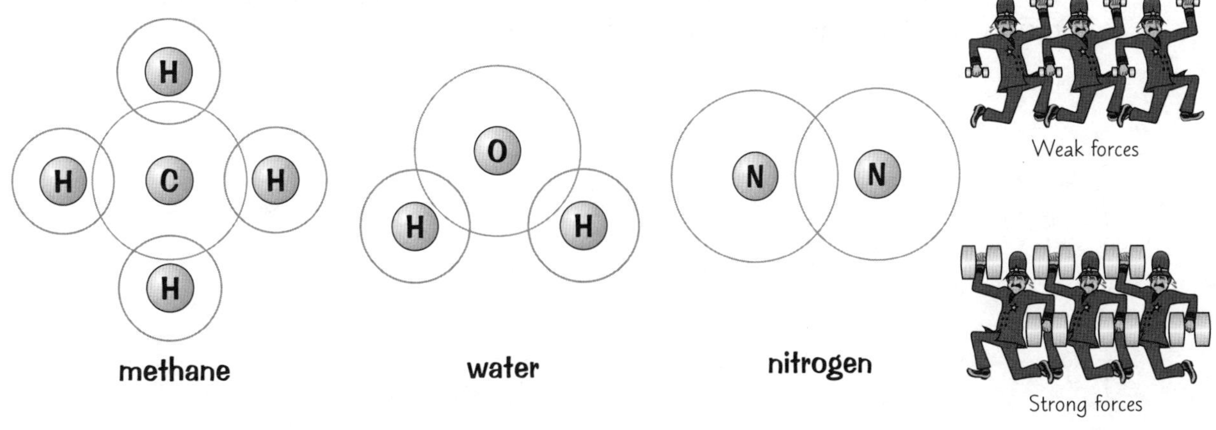

Q1 **Table 1** shows some properties of four substances.

Table 1

Substance	Melting Point (°C)	Conducts electricity when liquid?
A	1085	yes
B	1650	no
C	–39	yes
D	–102	no

a) Substance **A** is not a simple molecular substance.
How can you tell this using **Table 1**?

....

....

b) Which substance is a simple molecular substance? Explain your answer.

....

....

Q2 Phosphane and diphosphane are simple molecular substances. Their displayed formulas are shown in **Figure 1**.

Figure 1

phosphane

```
    H
    |
H — P — H
```

diphosphane

```
        H
        |
H — P — P — H
    |
    H
```

a) Phosphane has a similar structure and bonding to ammonia.
Draw a dot and cross diagram to represent the bonding in phosphane.
You only need to draw the outer shell electrons.

b) A chicken makes the following statement:

Diphosphane has a lower melting point than phosphane because it is a larger molecule than phosphane **cluck**.

Re-write the chicken's statement about diphosphane so that it is correct.

....

....

c) Phosphane is a gas at room temperature.
Which of the following statements explains why? Tick **one**.

☐ **A** The intermolecular forces between molecules of phosphane are strong.

☐ **B** The intermolecular forces between molecules of phosphane are weak.

☐ **C** The covalent bonds between atoms in phosphane are strong.

☐ **D** The covalent bonds between atoms in phosphane are weak.

Don't make me remember all this — ammonia little lad...

You've got to know what the properties of simple molecular substances are. If you can understand why simple molecular substances have these properties, it makes remembering them a lot easier. You'll be able to tackle any old question then.

Polymers and Giant Covalent Structures

Polymers are really repetitive — with enough practice, you should be an expert. Off you go...

Warm-Up

Polymers are long molecules. Covalent bonds hold the atoms in polymers together.
Small repeating units make up the long polymer.
Giant covalent structures are large networks of atoms which are all covalently bonded.

Complete the passage below by adding in the correct missing words.

intermolecular forces | solid | ionic | repeating units | covalent | liquid

Polymers are chains of ....

Polymers have stronger .... between them than simple .... molecules. This means they're usually .... at room temperature.

Q1 **Table 1** shows the formulas of three polymers.
Complete **Table 1** by adding the missing formulas.

Table 1

Name	Molecular formula of monomer	Molecular formula of polymer	Displayed formula of polymer
Poly(propene)	C_3H_6	....	$\left(\begin{array}{cc} H & H \\ \vert & \vert \\ -C & -C- \\ \vert & \vert \\ H & CH_3 \end{array} \right)_n$
PTFE	C_2F_4	$(C_2F_4)_n$	
Poly(styrene)	....	....	$\left(\begin{array}{cc} H & H \\ \vert & \vert \\ -C & -C- \\ \vert & \vert \\ C_6H_5 & H \end{array} \right)_n$

This symbol has the formula C_6H_5.

Q2 **Figure 1** shows part of the structure of substance **X**.

Figure 1

Substance **X** contains two types of atom.
Each atom forms four covalent bonds.

a) Predict the state of substance **X** at room temperature.

....

b) Tick the correct box to complete the following sentence.

Substance **X** must be made up of...

...two metals. ☐ ...two non-metals. ☐ ...a metal and a non-metal. ☐

Q3 **Table 2** shows the boiling points of silicon dioxide and carbon dioxide.

If something sublimes it turns straight from a solid into a gas.

Table 2

Compound	Boiling point (°C)
carbon dioxide	–78 (sublimes)
silicon dioxide	2230

Does silicon have the same symbol in Spain?

Si

a) Carbon dioxide and silicon dioxide are both covalently bonded substances. Tick the correct boxes to finish the sentences.

i) Carbon dioxide is a...

...simple molecular substance. ☐ ...giant covalent structure. ☐

ii) Silicon dioxide is a...

...simple molecular substance. ☐ ...giant covalent structure. ☐

b) What forces need to be overcome when boiling:

i) carbon dioxide?

....

ii) silicon dioxide?

....

c) Explain why silicon dioxide has a higher boiling point than carbon dioxide.

....

....

....

Polymers — they're really repetitive...

Some of these questions were a bit tricky and were meant to make you think a bit about what you know. If you're asked to explain melting or boiling temperatures, think about intermolecular forces and bonding between molecules or atoms.

Structures of Carbon

Diamonds are everyone's best friend. Other carbon structures are pretty nifty too. Have a look.

Warm-Up

Carbon atoms can arrange themselves into different structures. These include diamond, graphite, graphene and the various fullerenes, such as carbon nanotubes.

Each of these structures have different properties.
Complete this table showing the properties of several structures of carbon.

Name of structure	Description of structure	Conducts electricity?
....	giant covalent	no
....	layers of carbon atoms arranged in hexagons with no covalent bonds between layers	....
....	single layer of carbon atoms arranged in hexagons	....
....	molecules of carbon shaped like hollow balls or cylinders	cylindrical molecules conduct electricity

Q1 **Figure 1** shows a diagram of a battery. The electrodes make up part of the electrical circuit.

Figure 1

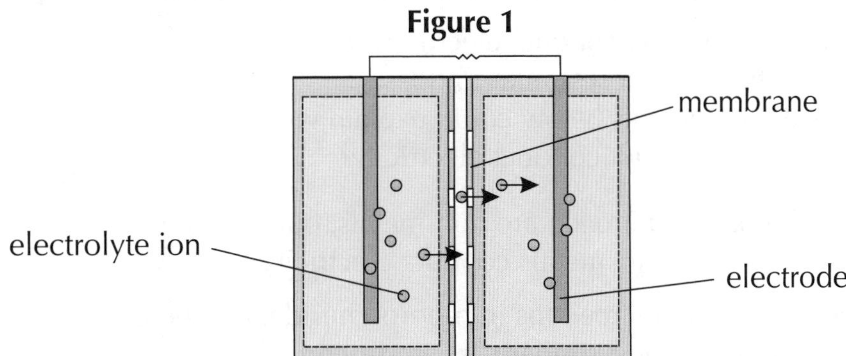

a) Why isn't diamond a suitable material for making electrodes?

....

b) The electrodes in some batteries are made from graphite.
These batteries can be improved by using metal electrodes which are coated with a thin layer of a different structure of carbon.

Suggest a structure of carbon that could be used to coat metal electrodes.

....

Q2 Structure **X** is made from carbon. Each carbon atom in structure **X** forms three covalent bonds. The atoms form a single sheet.

a) Name structure **X**.

....

b) Explain why structure **X** is strong.

....

....

Q3 Boron nitride is a compound which has several different structures. Some of these structures are similar to the different structures of carbon. This means they also have similar properties.

Figure 2

● boron ○ nitrogen

a) **Figure 2** shows the structure of c-boron nitride.

i) Which carbon structure is c-boron nitride is most similar to?

....

ii) Circle all of the expected properties of c-boron nitride.

hard **conducts electricity** **does not conduct electricity**

soft **high melting point** **low melting point**

b) **Figure 3** shows the structure of h-boron nitride.

i) Which carbon structure is h-boron nitride is most similar to?

....

Figure 3

● boron ○ nitrogen

ii) h-boron nitride doesn't conduct electricity, but the carbon structure it is most similar to does. Which of the statements below could explain why h-boron nitride does not conduct electricity?

☐ **A** Boron and nitrogen are non-metals, but carbon is a metal. Only metals conduct electricity.

☐ **B** There are no free ions in h-boron nitride, but there are in the most similar carbon structure.

☐ **C** There are no delocalised electrons in h-boron nitride, but there are in the most similar carbon structure.

☐ **D** h-boron nitride is an ionic compound. Ionic compounds only conduct electricity when they are molten or in solution.

My friend prefers oval cut diamonds. My personal favourite is Neil...

Diamond is pretty hard, but with plenty of practice, you should soon find questions on carbon structures pretty easy. Make sure you can recognise diagrams of the different structures and that you can explain their properties too.

Metallic Bonding

Seeing as you're (probably) such a fan, here are even more questions about bonding. Such a treat.

Warm-Up

Metals are elements which are good conductors. They are also solid at room temperature, and can be bent and shaped. Metals have a giant structure.

The outer shell electrons of metal atoms are free to move around the material — they are delocalised. The negatively charged electrons are strongly attracted to positively charged metal ions. These forces of attraction are called metallic bonds. This means the atoms in a metal are arranged in a regular pattern. The strong metallic bonds and delocalised electrons give metals many of their properties.

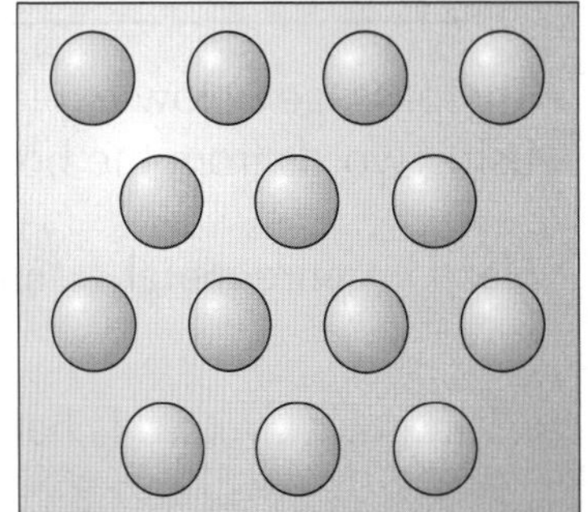

The diagram on the right shows the bonding in sodium metal. Complete the diagram by adding delocalised electrons and charges on the sodium ions.

Q1 Copper is often found in materials used to make gas pipes. These pipes need to be bent into the right shape.

Explain why the structure of pure copper means it can be easily bent.

....

....

Q2 Read the following statements about metals. Tick **all** the statements which are true.

- [ ] **A** Metals are suitable for insulating buildings to keep them warm.
- [ ] **B** Alloys contain regular layers of atoms which can slide over each other.
- [ ] **C** Metals conduct electricity because the electrons are held in fixed positions.
- [ ] **D** Pure metals are softer than alloys.

Q3 Household radiators are used to warm rooms. These are often made from metals.

a) Name the property of metals which makes them suitable for this purpose.

....

b) Explain why metals have this property.

....

....

Q4 Solder is an alloy of the metals lead and tin. Kuba is carrying out an experiment to compare the hardness of a piece of lead with the hardness of a piece of solder.

Kuba was given some instructions for the experiment, which are shown in **Figure 1**.

Figure 1

1. **Scratch the lead with a piece of solder.**
2. **Record your observations.**
3. **Scratch the solder with a piece of lead.**
4. **Record your observations.**
5. **The harder material should leave a mark on the softer material.**

The passage below explains the results of Kuba's experiment. Fill in the gaps in the passage using words from the box. You may use each word once, more than once, or not at all.

lead	pure metal	shape	solder	alloy	identical	outer electrons	different

.... is harder than .... because it's an ....

The atoms in solder are .... sizes so the layers of atoms lose

their .... The metal atoms in .... can't

slide over each other as easily as in ....

This means .... will mark ....

Q5 Some light bulbs contain a metal wire called a filament.
When electricity is passed through a filament, it gets hot and produces light.

I have a bright idea

I'm de-lighted to hear that

a) Light bulb filaments must reach high temperatures without becoming damaged.

i) Circle the property of metals which means they can do this.

hard **conducts electricity** **soft** **electrical insulator**
high melting point **thermal conductor** **thermal insulator**

Thermal insulators don't conduct heat. Electrical insulators don't conduct electricity.

ii) Explain why metals have this property.

....

....

....

....

b) Give one other property of metals which means they can be used as filaments.

....

Pure metal's the best music — mixed metal's really alloying...

There isn't loads to learn when it comes to metallic bonding. It's important to be able to relate properties of metals and alloys to the bonding within them though. And make sure you don't confuse metallic bonding with the other types...

States of Matter and Changing State

Like teleporting from Alaska to Texas — one moment you're frozen and the next you've melted...

Warm-Up

There are three states of matter — solids, liquids and gases. The state that a substance is in depends on how strong the forces of attraction between the particles are.

Particle theory is a model that can be used to describe the three states of matter.

In particle theory, particles are seen as small, solid spheres. The way that these particles behave is what makes solids, liquids and gases different.

Sketch the arrangement of particles in a solid, liquid and gas in the boxes below.
Draw each particle as a small circle.

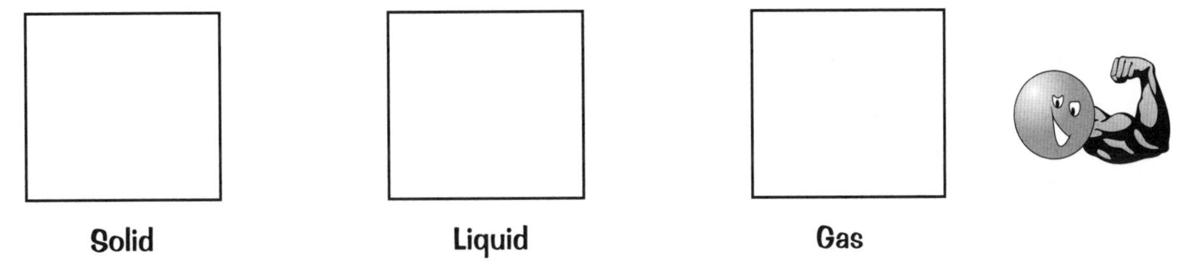

Q1 **Table 1** shows some information about a number of compounds. Fill in the empty row in **Table 1** with the correct state of each compound at room temperature (20 °C).

Table 1

Substance	Water	Sulfur	Mercury	Silica
Melting point (°C)	0	115	–39	1713
Boiling point (°C)	100	445	357	2950
State at room temperature	....	....	....	....

Q2 **Figure 1** shows an experiment in which sulfuric acid (H_2SO_4) is reacted with magnesium carbonate ($MgCO_3$).

Figure 1

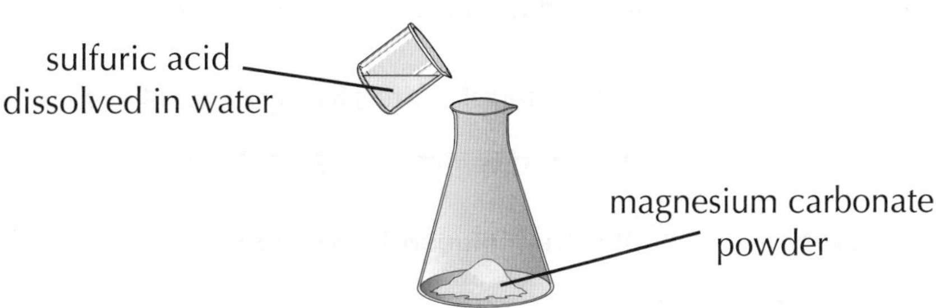

Use **Figure 1** to complete the equation below by adding in state symbols.

$$MgCO_{3\ldots\ldots\ldots} + H_2SO_{4\ldots\ldots\ldots} \rightarrow MgSO_{4(aq)} + CO_{2(g)} + H_2O_{(l)}$$

Q3 **Table 2** shows the melting and boiling points of three compounds. **Table 3** does not.

Table 2

Compound	Melting point (°C)	Boiling point (°C)
propane	–188	–42
propanol	–126	97
glycerol	18	290

Table 3

a) Which of the compounds in **Table 2** freezes at the highest temperature?

....

b) Which of the compounds in **Table 2** condenses at the lowest temperature?

....

c) **Table 4** lists some possible ways that the particles in the compounds in **Table 2** could be arranged.

i) In the middle column, write down the state that matches the arrangement of the particles.

ii) Use the temperatures in the box below to fill in the column on the right, to show a temperature at which the particles could have this arrangement. You may only use each temperature once.

–138 °C **–273 °C** **312 °C** **57 °C**

Use Table 2 to help you.

Table 4

Arrangement of particles	State	Temperature (°C)
Propanol particles are closely packed and regularly arranged.	....	....
Glycerol particles are spaced far apart.	....	....
Propanol particles take up a fixed volume but don't have a fixed shape.	....	....
Propane particles form a fixed shape and volume.	....	....

d) The passage below explains why glycerol and propane have different melting points. Circle the correct words in bold to complete the passage.

The forces between glycerol molecules are **stronger / weaker** than the forces between propane molecules. More energy is needed to overcome the forces between **glycerol / propane** molecules than **glycerol / propane** molecules.

Boil the kettle will you — set those particles free...

Particle theory is likely to crop up a lot in chemistry, so it's dead important you understand it all really well. And make sure you can interpret melting and boiling points to work out the state of a substance at a certain temperature.

Relative Formula Mass

Time for a little bit of maths, but don't let that put you off...

Warm-Up

The <u>relative formula mass</u> (M_r) of a compound is calculated by <u>adding</u> together the <u>relative atomic masses</u> (A_r) of all the <u>atoms</u> in the compound's <u>molecular formula</u>.

Complete the calculation to show how you would calculate the relative formula mass of magnesium hydroxide, $Mg(OH)_2$. Relative atomic masses: Mg = 24, O = 16, H = 1

Relative formula mass = ... + [(16 + ...) × ....] = <u>58</u>

Q1 Element **X** reacts with water.
The compound formed in the reaction has the formula XOH.

A positive reaction

a) XOH has a relative formula mass of 56.
The relative atomic mass of H = 1 and of O = 16.
Calculate the relative atomic mass (A_r) of element **X**.

relative atomic mass (A_r) = ...

b) What is element **X**?

....

Q2 Ammonium nitrate is a compound made up of nitrogen, hydrogen and oxygen.
It has the formula NH_4NO_3.

a) Calculate the relative formula mass (M_r) of ammonium nitrate.
Relative atomic masses (A_r): N = 14, O = 16, H = 1

Relative formula mass = ...

b) Calculate the percentage mass of oxygen in ammonium nitrate. Use the equation:

$$\text{Percentage mass of element in a compound} = \frac{A_r \text{ of element} \times \text{number of atoms of element}}{M_r \text{ of compound}} \times 100$$

Percentage mass of oxygen = ... %

My Grandad makes great mash — he's got a special formula for it...

Remember, if a formula has brackets in it, it means the numbers of ALL the atoms inside the brackets need to be multiplied by the number that comes afterwards. For example, in $Al(OH)_3$ there are 3 oxygen atoms and 3 hydrogen atoms.

Conservation of Mass

The atoms in products don't just appear from nowhere — they ALL come from the reactants.

Warm-Up

During a chemical reaction no atoms are made or lost.
So no mass is gained or lost — mass is conserved.

Look at the equation below.
It shows magnesium oxide and water reacting together to form magnesium hydroxide.

$$MgO + H_2O \rightarrow Mg(OH)_2$$

Fill in the table to show how many of each type of atom there are in the reactants and in the products. The first row has been done for you.

Type of atom	Number of each type of atom	
	in reactants	in products
Mg	1	1
O	....	....
H	....	....

A conversation of mass

What do you notice about the numbers of atoms on each side?

....

Q1 The balanced symbol equation below shows a reaction between sodium and water.

$$2Na + 2H_2O \rightarrow 2NaOH + H_2$$

Relative atomic masses (A_r): Na = 23, H = 1, O = 16

a) i) Calculate the total relative mass of the reactants in this reaction.

total relative mass of reactants = ...

ii) Calculate the total relative mass of the products in this reaction.

total relative mass of products = ...

b) Explain how your calculations show that mass has been conserved in the reaction.

....

....

Q2 When iron wool burns, the iron reacts with oxygen in the air. This forms iron oxide. No other products are made.

4.4 g of iron wool is burnt. 6.3 g of iron oxide is produced. What mass of oxygen did the iron react with?

mass of oxygen = ... g

Q3 Lithium hydroxide reacts with nitric acid to form lithium nitrate and water.

8.0 g of lithium hydroxide reacts with 21.0 g of nitric acid. 6.0 g of water is formed. Calculate the mass of lithium nitrate formed.

mass of lithium nitrate = ... g

PRACTICAL

Q4 A scientist added 6 g of zinc carbonate and 53 g of dilute sulfuric acid to a strong conical flask. The equation below shows the reaction that took place:

$$ZnCO_{3(s)} + H_2SO_{4(aq)} \rightarrow ZnSO_{4(aq)} + CO_{2(g)} + H_2O_{(l)}$$

When the reaction finished, the scientist measured the total mass of the products in the flask. She repeated the experiment. **Table 1** shows the results for both experiments.

Table 1

Experiment	Total mass of reactants / g	Total mass of products / g
1	59	59
2	59	57

In one experiment, the scientist sealed the conical flask when she added the reactants. In the other experiment, she left the conical flask open.

a) Use **Table 1** to identify whether the flask was open in experiment **1** or experiment **2**.

....

b) Explain your answer.

....

....

....

....

Eating biscuits while revising helps me conserve my mass...

It doesn't matter if you're looking at the relative formula masses or the actual reacting masses — mass is always conserved in reactions. If the mass changes, it could be because one of the products is a gas that has escaped from the reaction vessel.

Concentrations of Solutions

Time for some concentration and even more maths. You're in for a treat...

Warm-Up

Concentration is a measure of the mass of a substance dissolved in a certain volume of solution. The substance that is dissolved is called the solute.

The more substance that has dissolved in a certain volume of solution, the more concentrated the solution.

Complete the formula that you would you use to calculate the concentration of a solution.

.... = $\dfrac{\text{mass of dissolved substance}}{\text{..}}$

Bernard always checks his solutions carefully

Q1 Ammonium sulfate, $(NH_4)_2SO_4$, is a soluble salt. It is often used as a fertiliser.

Erin is preparing a solution of ammonium sulfate.
She dissolves 2.7 g of ammonium sulfate in 1.8 dm^3 of water.
What is the concentration of the solution in g/dm^3?

concentration = ... g/dm^3

Q2 Sujit makes a 250 cm^3 solution of copper sulfate.
The solution has a concentration of 32 g/dm^3.

You'll need to convert 250 cm^3 into dm^3 first.

What mass of copper sulfate does Sujit's solution contain?

mass = ... g

You'll need all your concentration to find these solutions...

Make sure you're super-happy using and rearranging the formula for concentration.
Practice makes perfect, and if you find formula triangles helpful, you might want to use one here.

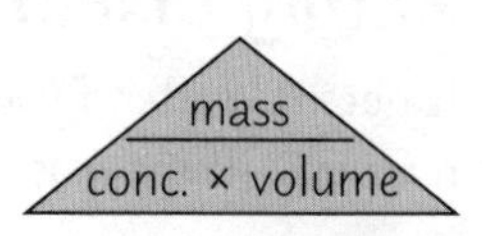

Acids and Bases

Acids and bases are pretty key to this section, so you had better get to grips with what they are.

Warm-Up

Acids have a pH which is less than 7. Bases have a pH which is bigger than 7.
Alkalis are bases which dissolve in water.

Which of the following is the correct equation for a neutralisation reaction?

☐ A $H^+_{(aq)} + OH^-_{(aq)} \rightarrow H_2O_{(l)}$

☐ B $H^-_{(aq)} + OH^+_{(aq)} \rightarrow H_2O_{(l)}$

☐ C $H^+_{(aq)} + H_2O_{(l)} \rightarrow OH^-_{(aq)}$

☐ D $OH^-_{(aq)} + H_2O_{(l)} \rightarrow H^+_{(aq)}$

Acid bass

Q1 Ant stings hurt because they contain formic acid.
Table 1 shows the pH of different acids and bases.

Table 1

Substance	pH
lemon juice	3
baking soda	9
milk	6

a) Which substance in **Table 1** could be rubbed on an ant sting to stop it from hurting?

....

b) Name a piece of apparatus that could be used to measure the pH of formic acid.

....

Q2 Complete the sentences below by circling the correct word in bold.

When an acid is mixed with a base a **neutralisation / oxidation** reaction takes place.

Hydrogen ions in the **acid / base** react with hydroxide ions in the **acid / base**.

This reaction forms **carbon dioxide / water**.

After all of the acid and the base have reacted, the solution has a pH of **seven / fourteen**.

Know the pH scale — don't make basic errors...

Universal indicator can be used to measure pH. It gradually changes colour as the pH changes. A solution with a pH of 0 is red, and changes to orange and yellow as the pH increases. It is green when it's neutral, changing to blue and then purple as the pH reaches 14. Much like a rainbow really — "red and orange and yellow and green...".

Reactions of Acids

My first reaction to these questions was a long sigh. But you just have to crack on sometimes.

Warm-Up

Acids react with bases and alkalis in neutralisation reactions.

Metal oxides and hydroxides are bases which react with acids to form...

...a salt and carbon dioxide ☐ ...a salt and water ☐

You can make soluble salts in the lab by reacting an acid with an insoluble base.

Q1 Draw lines to join the salt to the correct acid and base that can be used to make it. Each acid can be used more than once. One has been done for you.

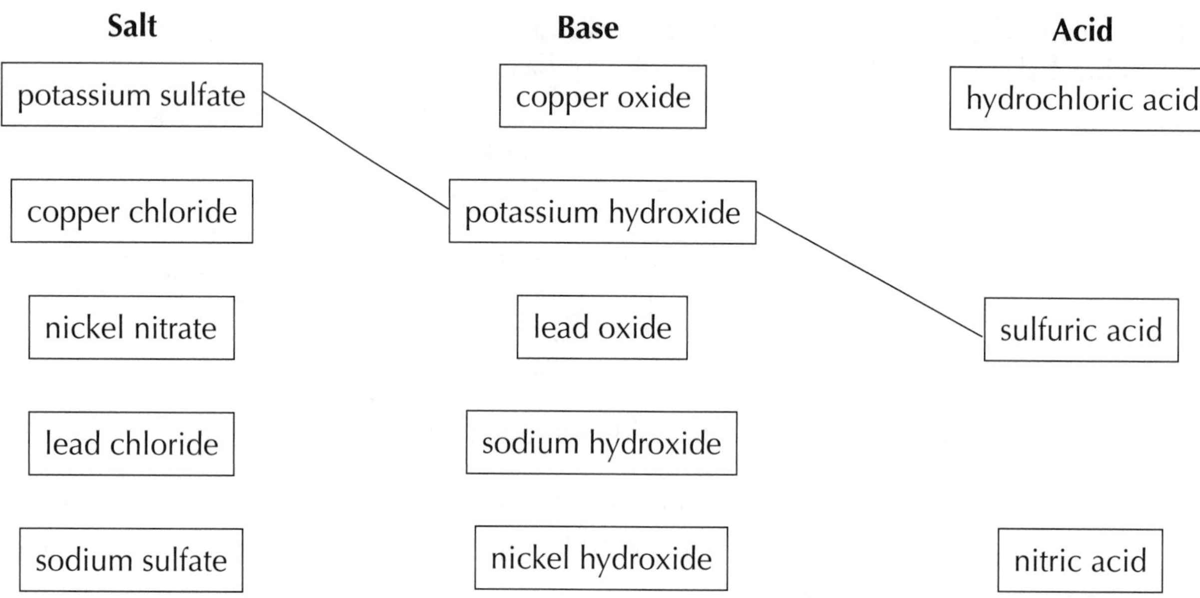

Q2 The symbol equation for the neutralisation reaction below is not balanced.

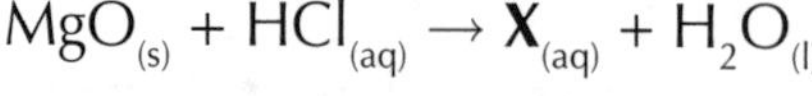

$$MgO_{(s)} + HCl_{(aq)} \rightarrow \mathbf{X}_{(aq)} + H_2O_{(l)}$$

a) Circle the acid in the equation above.

b) **X** is a mystery product. What is the formula of **X**? Tick **one**.

☐ **A** $Mg(OH)_2$

☐ **B** $Mg(ClO)_2$

☐ **C** $MgCl_2$

☐ **D** MgH_2

c) What salt would be formed if sulfuric acid reacted with magnesium oxide?

....

PRACTICAL

Q3 Leona is making zinc sulfate, using an acid and an insoluble base.

a) Which of the following is an acid-base pair that Leona could use to make zinc sulfate?

☐ **A** zinc chloride and hydrochloric acid

☐ **B** zinc hydroxide and hydrochloric acid

☐ **C** zinc and sulfuric acid

☐ **D** zinc oxide and sulfuric acid

Zinc sulfate is a soluble salt.

b) The boxes below show the steps involved in making zinc sulfate in the lab. Write the numbers 1-6 in the boxes to put the steps in the correct order.

Filter the unreacted base from the solution of the salt. ☐

Add insoluble base until no more reacts. ☐

Warm dilute acid using a Bunsen burner. ☐

Leave the solution to cool and crystallise. ☐

Heat the salt solution over a water bath to evaporate some solution. ☐

Filter the crystals and dry them. ☐

Q4 Amir has an unknown copper salt. **Table 1** shows the possible identities of the salt.

Table 1

Possible salt identity	Formula
copper carbonate	$CuCO_3$
copper hydroxide	$Cu(OH)_2$

Amir mixes his copper salt in a flask with some dilute hydrochloric acid. When the salt is added, the solution fizzes as a gas is produced.

Amir says that his salt must be copper carbonate. Is Amir correct? Explain how you can tell.

Think about what products the two salts would make when they react with acid.

....

....

....

....

Phhheeww — the right reaction to finishing this page...

There's quite a lot to learn in this section. Make sure you can predict the salt that will form when an acid reacts with a base. You also need to know how you use acids and bases to go about making soluble salts in the lab too.

The Reactivity Series and Extracting Metals

Oh man, this one is my favourite series. You won't be able to find it on a box set though...

Warm-Up

The reactivity series is a list of metals in order of their reactivity.

The higher up the reactivity series a metal is, the more easily it forms...

...a negative ion. ☐ ...a positive ion. ☐ ...a heavy metal band. ☐

Which two non-metals are often included in the reactivity series?

1. ....

2. ....

Most metals react with oxygen, forming oxides. These oxides are often ores that the metals need to be extracted from. Some metals are very unreactive so they are found in the earth as the metal themselves.

Give an example of a metal found in the earth as itself.

....

Q1 Complete the following passage using the words in the box below. You don't need to use every word.

electrolysis	reduction	oxidation	more	below	above	less

Carbon can be used to extract metals that are .... it in the reactivity series.

Oxygen is removed from the metal oxide in a process called ....

Carbon cannot be used to extract metals which are .... reactive than it.

These metals must be extracted using other methods, such as ....

Q2 Caveman Trev dropped some copper ore onto a wood fire. When the fire burned out, copper metal was left behind.

Wood contains carbon.

Which of the following explains why copper was left behind? Tick **one**.

☐ **A** Carbon is more reactive than copper.
Carbon reduces copper ore by removing oxygen from it.

☐ **B** Carbon is less reactive than copper.
Carbon reduces copper by removing oxygen from it.

☐ **C** Carbon is less reactive than copper.
Carbon oxidises copper by removing oxygen from it.

☐ **D** Carbon is more reactive than copper.
Carbon oxidises copper by removing oxygen from it.

Q3 Aluminium metal can be used to extract iron from iron oxide, Fe_2O_3.
This process forms aluminium oxide, Al_2O_3, and iron.
The balanced equation for the reaction that takes place is shown below:

$$Fe_2O_3 + 2Al \rightarrow Al_2O_3 + 2Fe$$

a) i) Which element or compound is oxidised in this reaction?

....

ii) How can you tell?

....

b) Kamran heated some aluminium oxide with carbon. Aluminium was not extracted. What does this tell you about the reactivity of aluminium?

....

Q4 Imagine three new metals have been discovered.
Table 1 shows how these metals are extracted.

Table 1

Metal	How is it extracted?
antium	Found as the metal itself.
bodium	Can't be extracted by reduction with carbon.
kandium	By reducing kandium oxide with carbon.

a) Use the information in **Table 1** to put these metals in order of reactivity.

REACTIVITY ↑

1. ....

2. ....

3. ....

b) The equation below shows the reduction of kandium oxide.

Imagine that kandium has the symbol 'Ka'.

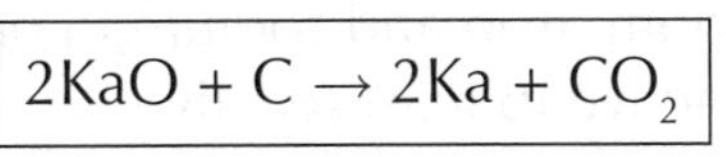

$$2KaO + C \rightarrow 2Ka + CO_2$$

What happens to kandium oxide when it is reduced by carbon?

....

Iron and oxygen were inseparable friends — until carbon came along...

It's not the most thrilling thing in this book but make sure you understand the reactivity series. Remember, a metal's position in the reactivity series tells you how reactive it is compared to other metals, as well as how it can be extracted.

Reactions of Metals

Here's a heads up before you start. Sodium reacts violently with water — it got three years for a salt.

Warm-Up

Metals react with acids to form...

...a salt and water. ☐ ...hydrogen and a salt. ☐

Some metals also react with water, forming a metal hydroxide and...

...hydrogen. ☐ ...water. ☐ ...carbon dioxide. ☐

Displacement reactions involve one metal kicking another one out of a compound.

A metal will displace a **less / more** reactive metal from a compound.

Q1 **Figure 1** shows a reactivity series. Use **Figure 1** to decide which of the displacement reactions below are possible. Tick **two**.

☐ **A** zinc sulfate + iron → iron sulfate + zinc

☐ **B** magnesium sulfate + zinc → zinc sulfate + magnesium

☐ **C** iron + copper nitrate → iron nitrate + copper

☐ **D** magnesium + iron chloride → magnesium chloride + iron

☐ **E** magnesium + lithium chloride → magnesium chloride + lithium

Figure 1

Li	lithium
Mg	magnesium
Zn	zinc
Fe	iron
Cu	copper

REACTIVITY ↑

Q2 Samira reacted some metals with cold water. She recorded her observations in **Table 1**.

Table 1

Metal added to water	Observation
Sodium	Vigorous reaction with lots of bubbles produced. Sodium disappears.
Iron	No bubbles. No change to iron.
Calcium	Bubbles produced. Calcium disappears.

a) Use **Table 1** to put calcium, iron and sodium in order of reactivity from most reactive to least reactive.

1. ....

2. ....

3. ....

b) Samira says that when she added iron to water, iron oxide and hydrogen would have formed straight away. Do you agree with Samira? Explain your answer.

....

....

PRACTICAL

Q3 Jenson compared the reactivity of three metals — **K**, **L** and **M**.
He added these metals into separate beakers which contained hydrochloric acid.

He then measured the change in temperature over the first 150 seconds of the reaction.
Figure 2 shows Jenson's results.

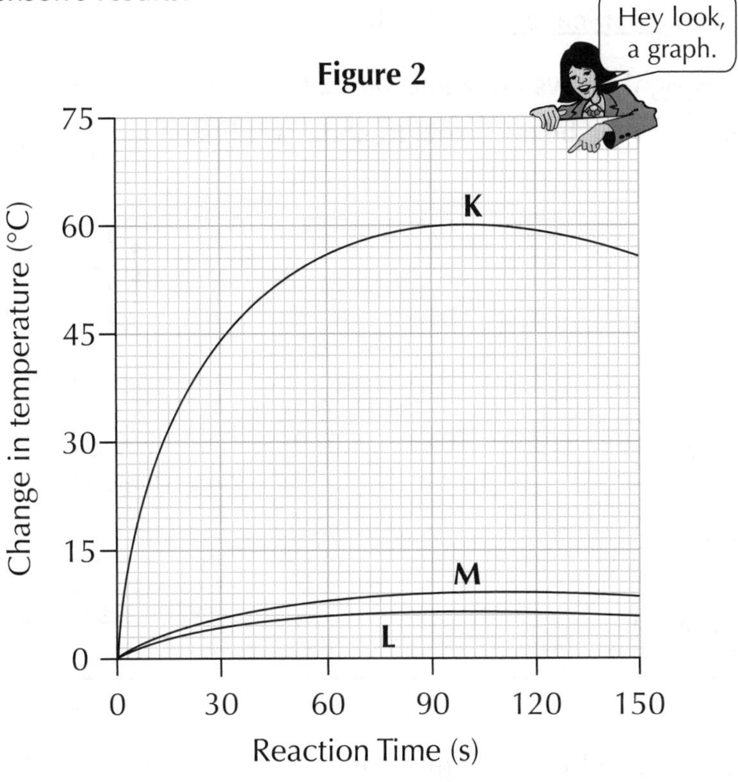

Figure 2

a) Write the three metals, **K**, **L** and **M**, in order, from most reactive to least reactive.

Most reactive .... Least reactive

b) Which of the following is the independent variable in Jenson's experiment?

- ☐ **A** Reaction time
- ☐ **B** Change in temperature
- ☐ **C** The type of metal used in the reaction
- ☐ **D** The type of acid used

c) Give **two** variables that need to be kept the same for this experiment to be a fair test.

1. ....

2. ....

d) Could metal **M** displace metal **K** from a compound of **K**? Explain your answer.

....

....

Compare your friends' reactivity by pranking them...

You might be given the results of an experiment in which metals are reacted with an acid or water. You'll have to use the results to put the metals in order of reactivity. More reactive metals will react more vigorously than less reactive metals.

 ☐  ☐  ☐

Electrolysis

Passing a current through me would probably cause a reaction. It's pretty similar with salts too...

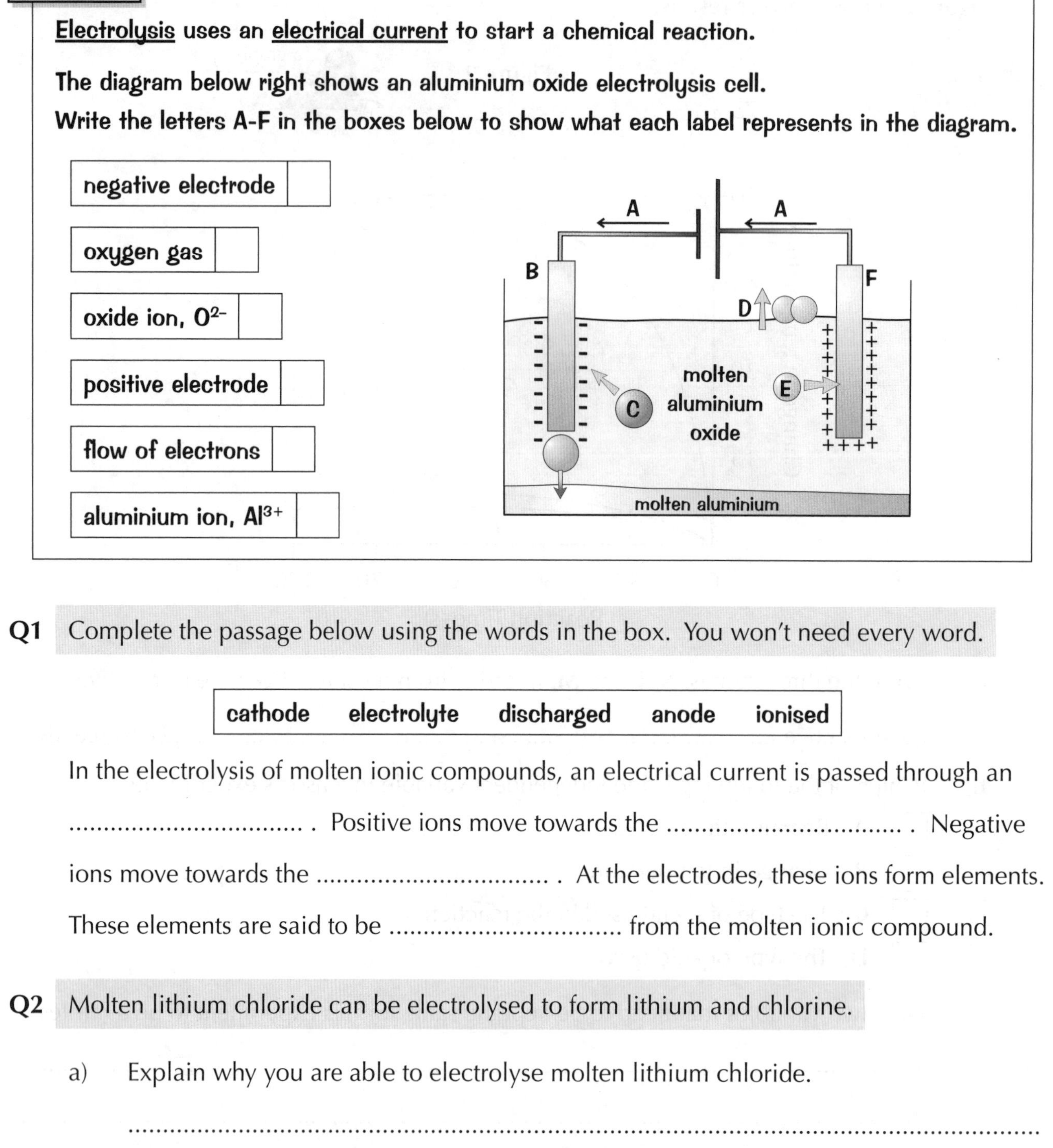

Warm-Up

Electrolysis uses an electrical current to start a chemical reaction.

The diagram below right shows an aluminium oxide electrolysis cell.
Write the letters A-F in the boxes below to show what each label represents in the diagram.

negative electrode	
oxygen gas	
oxide ion, O^{2-}	
positive electrode	
flow of electrons	
aluminium ion, Al^{3+}	

Q1 Complete the passage below using the words in the box. You won't need every word.

cathode	electrolyte	discharged	anode	ionised

In the electrolysis of molten ionic compounds, an electrical current is passed through an .... Positive ions move towards the .... Negative ions move towards the .... At the electrodes, these ions form elements. These elements are said to be .... from the molten ionic compound.

Q2 Molten lithium chloride can be electrolysed to form lithium and chlorine.

a) Explain why you are able to electrolyse molten lithium chloride.

....

....

b) When molten lithium chloride is electrolysed, which element forms at the:

i) anode?

....

ii) cathode?

....

Q3 A company has built a new plant which extracts aluminium from its ore using electrolysis.

a) Explain why the company melts the aluminium oxide ore before electrolysing it.

....

....

....

b) Aluminium oxide has a high melting point. Lots of energy is needed to melt it.
What can the company do to reduce the cost of heating aluminium oxide?

....

c) The company often has to replace one of the electrodes.
Which electrode must they replace? Explain why they must replace this electrode.

....

....

Q4 Ffion is carrying out the electrolysis of concentrated aqueous potassium bromide.

a) Which of the following ions are found in solution when concentrated potassium bromide is dissolved in water? Circle **four**.

H_2 Br^- H^- OH^- K^+ H^+ OH^+

b) i) Which element is produced at the anode when concentrated aqueous potassium bromide is electrolysed? Tick **one**.

- [ ] **A** bromine
- [ ] **B** hydrogen
- [ ] **C** potassium
- [ ] **D** oxygen

ii) Explain your answer.

....

c) i) Which element is produced at the cathode when concentrated aqueous potassium bromide is electrolysed? Tick **one**.

- [ ] **A** bromine
- [ ] **B** hydrogen
- [ ] **C** potassium
- [ ] **D** oxygen

ii) Explain your answer.

....

Electrolysis — a very current topic...

If you find it hard remembering which electrode is which, just think of PANCakes — Positive Anode, Negative Cathode. When aqueous solutions are electrolysed, remember that the ions formed by water can move towards the electrodes too.

Exothermic and Endothermic Reactions

Time to practise stuff on energy changes in chemical reactions. I bet you're dead excited...

Warm-Up

Exothermic reactions give out energy to the surroundings, usually in the form of heat.
This causes the temperature of the surroundings to rise.
Endothermic reactions take in energy from the surroundings.
This causes the temperature of the surroundings to fall.

Draw lines to match the reactions or processes below to the correct label on the right.

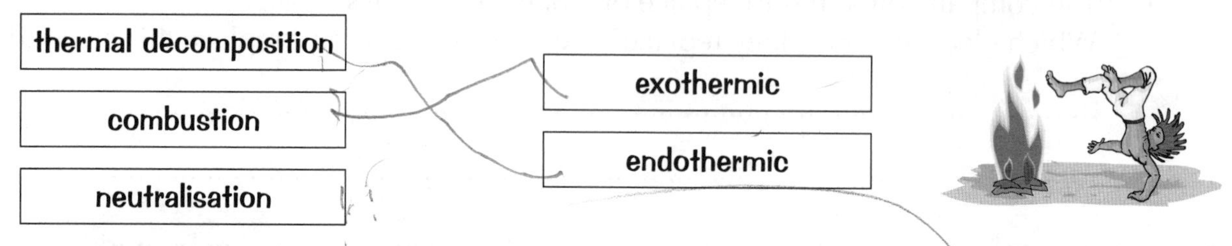

PRACTICAL

Q1 Lorna and Kezia are investigating the temperature change during a neutralisation reaction.

They each add 50 cm^2 of an alkaline solution to 25 cm^2 of an acid solution, then measure the temperature of the mixture every two seconds.
Figure 1 shows their experimental set-ups.

Figure 1

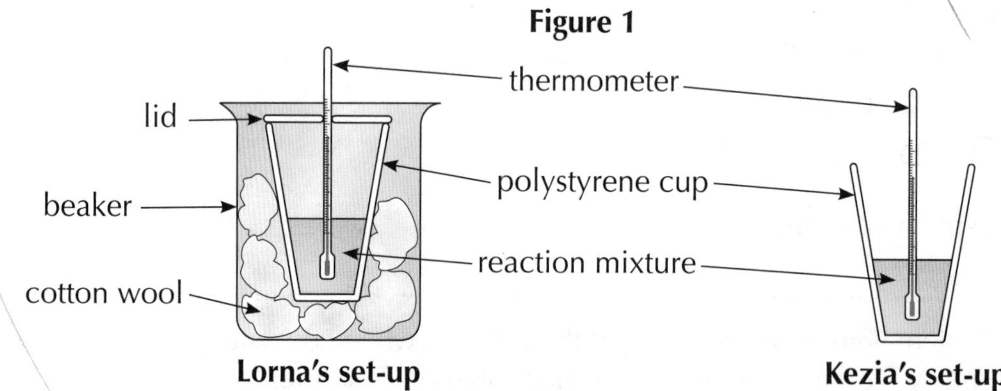

a) Kezia thinks that the reaction will cause the temperature of the mixture to rise. Lorna thinks that the temperature will fall. Who is correct?

Kejza is correct

b) Using **Figure 1**, explain why Lorna is likely to have more accurate results than Kezia.

because the conton wool acts as a insulator and stops heat escaping

Endothermic reactions — they're pretty cool...

Energy can't be made or destroyed. So the amount of energy in the universe will be the same before and after a reaction. But energy can move around — the products of a reaction have either taken energy in or transferred it to the surroundings.

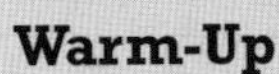

Reaction Profiles

If you thought you were done with endothermic and exothermic reactions, you were mistaken...

Warm-Up

Chemical reactions only take place if the reacting particles collide with enough energy. What name is given to the minimum amount of energy needed for a reaction to start?

....

Reaction profiles show the difference between the energies of the reactants and products in a reaction.

On the reaction profile to the right, circle the arrow which shows the total energy given out by the reaction.

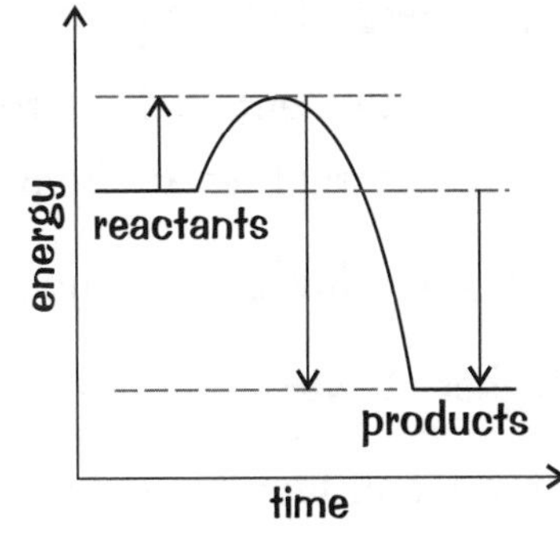

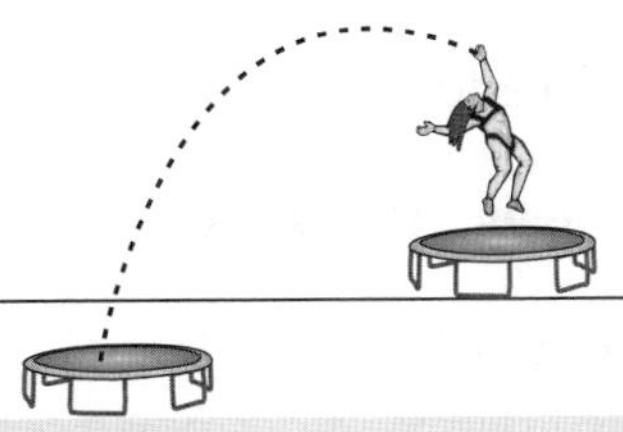

Q1 The reaction profiles in **Figure 1** show the energy changes in five chemical reactions.

For parts a) – d), write the letter of the graph(s) matching the description. Assume all the axes have the same scale.

a) An exothermic reaction.

....

b) An endothermic reaction.

....

c) The reaction with the largest activation energy.

....

d) The reaction which needs the smallest amount of energy to start.

....

Figure 1

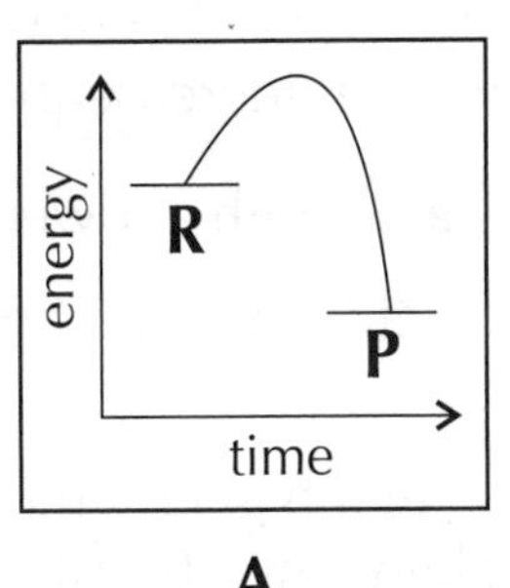

A

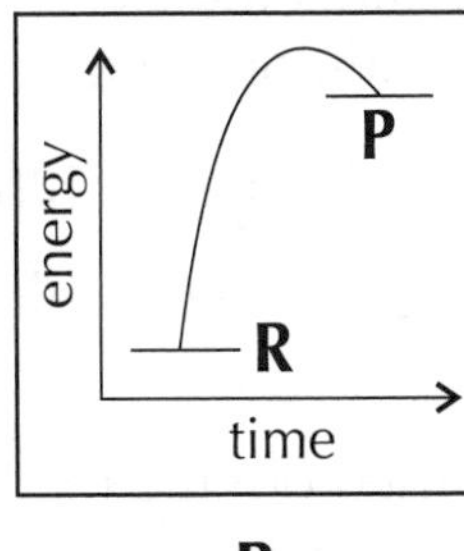

B

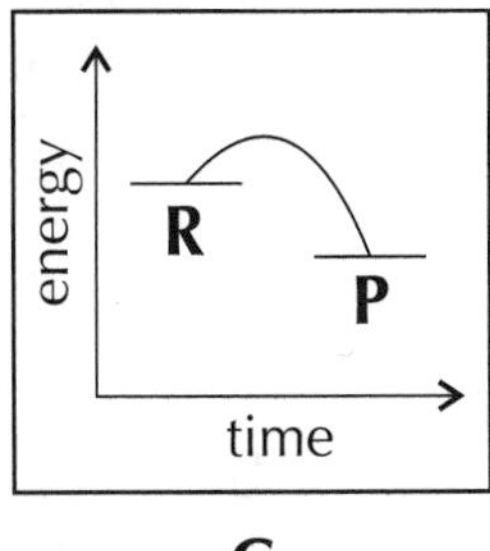

C

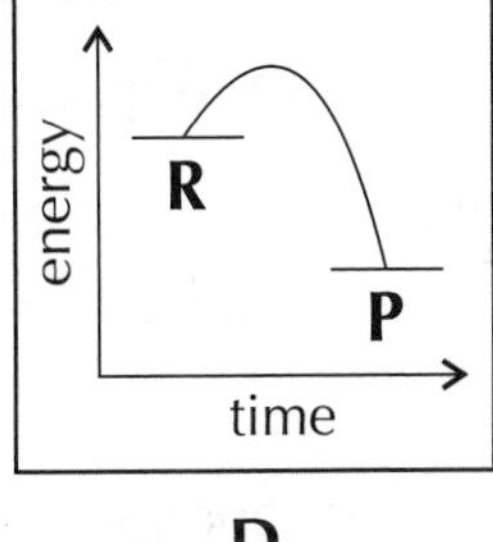

D

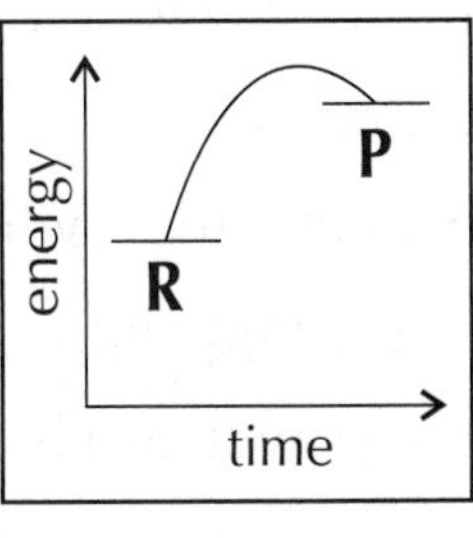

E

Key: **R** = reactants, **P** = products

A nice mug of coffee — that's my morning activation energy...

This stuff shouldn't be too tough but it can be easy to get caught out. If you get a little muddled, practice drawing out reaction profiles for endothermic and exothermic reactions. Make sure you can label the activation energy on them too.

Rates of Reaction

Get your skates on — this section's all about speed. Well, sort of anyway...

Warm-Up

The rate of reaction is how fast reactants are used up and products are made.

The rate is affected by how often particles collide.
The rate of reaction also depends on particles colliding with enough energy to react. This is called activation energy.

Boo!

Okay.

The four statements below are about rates of reaction.
Circle the correct words from each pair to complete the sentences.

The **higher / lower** the temperature, the faster the rate of reaction.

A **higher / lower** concentration or pressure will reduce the rate of reaction.

A smaller surface area of solid reactants **increases / decreases** the rate of reaction.

A catalyst **does / doesn't** change the rate of reaction.

Q1 Luisa is investigating the rate of reaction between 50 cm^3 hydrochloric acid and a 3 cm length of magnesium ribbon.

a) Luisa repeated the experiment with different conditions.
Which conditions would cause an increase in reaction rate? Tick **one** box.

- [ ] **A** Using hydrochloric acid with a lower concentration.
- [ ] **B** Using hydrochloric acid with a higher concentration.

b) Luisa carried out the reaction at a higher temperature. The rate of reaction increased.
The following sentences explain why. Tick the boxes to finish the sentences.

When the temperature is increased, the particles all move...

...more slowly. [ ] ...more quickly. [ ] ...at the same speed. [ ]

This means that they will collide...

...more frequently. [ ] ...less frequently. [ ] ...just as often. [ ]

The particles will also have...

...more energy. [ ] ...less energy. [ ] ...the same amount of energy. [ ]

So more of the particles will have enough energy to make the reaction happen.

c) The equation for the reaction is:

magnesium + hydrochloric acid → magnesium chloride + hydrogen

Would the reaction equation change if a catalyst was used in the reaction?

....

PRACTICAL

Q2 Saz reacted together 5 g of marble chips and 100 cm^3 of hydrochloric acid. She measured the volume of carbon dioxide produced. She then repeated the experiment at a different temperature. **Table 1** shows the results of both reactions.

Table 1

Time (s)	Volume of CO_2 (cm^3)	
	Reaction 1	Reaction 2
10	14	24
20	25	42
30	36	57
40	46	69
50	54	77
60	62	80
70	70	80
80	76	80
90	80	80
100	80	80

Figure 1

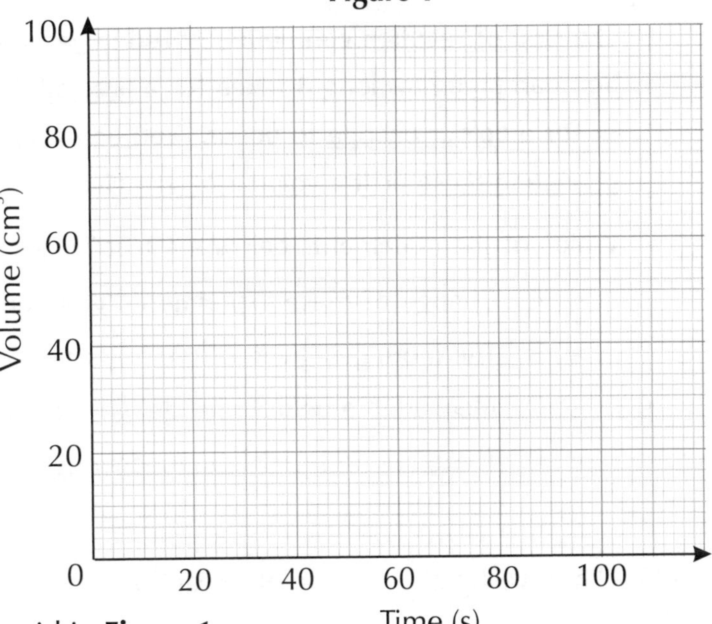

a) Plot the data from **Table 1** onto the grid in **Figure 1**. Draw each reaction as a separate curve.

b) Which reaction is faster?

....

c) Saz changed the temperature of the second reaction to change the rate of reaction. Suggest another factor which Saz could have changed to change the rate of reaction.

....

Q3 Ivona gets lost while exploring in the jungles of Cumbria. She decides to light a fire using her lighter and some large chunks of wood that she finds. The wood burns in a combustion reaction with oxygen in the air.

The passage below explains why the fire will burn more quickly if she chops the wood into smaller chunks before burning it. Use words from the box to fill in the gaps. You don't need to use all of the words.

inside	surface	reactions	more	smaller	fewer	bigger	collisions

The fire will burn more quickly with smaller chunks of wood because they have a .... surface area than a large piece of wood. This means that .... oxygen molecules can collide with the .... of the wood, increasing the chance of successful collisions between reacting particles.

Collision theory has its uses — you'll never be cold in a jungle again...

Changing certain factors will affect how often particles collide. This means the rate of reaction will also change. Don't get caught out by thinking catalysts affect the collision frequency though — they're too cool for that.

PRACTICAL Measuring Rates of Reaction

This section's all about measuring speeds. Like a speed camera, but loads more exciting. Super.

Warm-Up

The rate of reaction can be measured in several ways. You can time how long it takes for a precipitate to form, or for a colour change to occur. You can also measure the change in mass or the volume of gas given off at regular intervals.

Which of the statements below are true? Tick one box.

- ☐ **A** Change-of-mass experiments give very subjective results.
- ☐ **B** The more gas given off during a given time interval, the faster the reaction.
- ☐ **C** Measuring a colour change with the naked eye gives very accurate results.
- ☐ **D** There are no hazards in measuring the volume of gas given off in an experiment.

Q1 Em investigated the rate of a reaction between marble chips and hydrochloric acid.

She calculated the change in the mass of the reaction mixture from the start of the reaction. She then repeated the experiment with a different concentration of hydrochloric acid. She kept everything else the same. **Figure 1** shows the results.

Figure 1

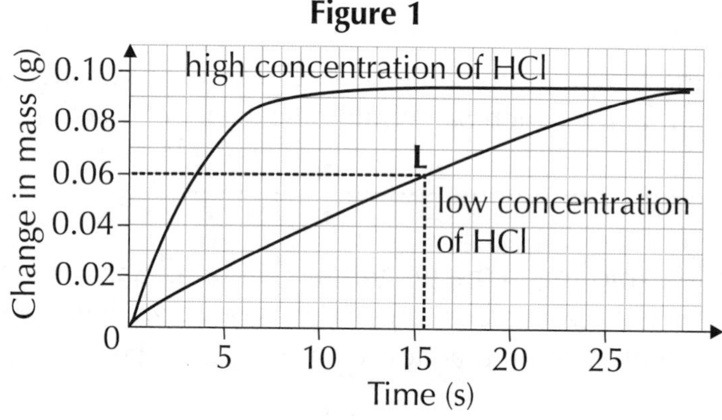

a) What valid conclusion can Em drawn from **Figure 1** about the reaction between hydrochloric acid and marble chips? Tick **one** box.

- ☐ **A** The reaction rate depends on the temperature of the reactants.
- ☐ **B** Increasing the concentration of the acid has no effect on the rate of reaction.
- ☐ **C** The reaction rate depends on the acid concentration.
- ☐ **D** The reaction rate depends on the mass of the marble chips used.

b) Calculate the mean rate of reaction between the origin and point **L** on the graph. Use the formula below. Give your answer to one significant figure.

$$\text{mean rate of reaction} = \frac{\text{change in mass (g)}}{\text{time taken (s)}}$$

Mean rate = ... g/s

Q2 Ari reacted 5 g of calcium with five different concentrations of sulfuric acid. She measured the volume of gas produced during the first minute of each reaction and repeated each measurement three times. **Table 1** shows her results.

Table 1

Sulfuric acid concentration (mol/dm^3)	Volume of gas produced (cm^3)			Mean volume of gas produced (cm^3)
	Experiment 1	Experiment 2	Experiment 3	
2.0	92	96	93	93.7
1.5	63	65	65	....
1.0	44	47	31	45.5
0.5	20	22	21	21.0

a) Circle the anomalous result in **Table 1**.

b) Complete the final column in **Table 1**.
Give your answer to one decimal place.

c) Circle the concentration of sulfuric acid in **Table 1** that produced the fastest rate of reaction.

Acid concentration

d) Suggest what may have caused the anomalous result.

....

Q3 Yasmin investigates the effect of temperature on the rate of the reaction between sodium thiosulfate solution and hydrochloric acid. She mixes the reactants together in a flask and times how long a cross placed under the flask takes to disappear. **Table 2** shows the results from her investigation.

Table 2

Temperature (°C)	20	30	40	50	60
Time taken for cross to disappear (s)	201	177		112	82

a) One of the values in the table is missing. Tick the most likely value for it.

☐ **A** 145 s ☐ **B** 192 s ☐ **C** 15 s

b) Give the dependent variable in Yasmin's experiment.

....

Anomalous — that's an odd word...

You should be a whizz at this stuff now. It's all data and tables and numbers — sounds like maths. Bubbles, gases and colour changes in chemistry hopefully make you more excited than maths though. I think.... maybe.... not.... um, okay.

Reversible Reactions

Some questions about going backwards now — it'd be great if this page beeped to warn you...

Warm-Up

Some reactions can go backwards — the <u>products</u> can react with each other to re-form the <u>reactants</u>. When the <u>concentrations</u> of products and reactants <u>stop changing</u>, a reversible reaction is at <u>equilibrium</u>.

Let's do that scene again.

Na Cl

A rehearsable reaction

Tick any of the statements below that are <u>false</u>.

- ☐ **A** A reaction at equilibrium will only be going in one direction.
- ☐ **B** Endothermic reactions are always reversible.
- ☐ **C** Equilibrium is the point at which the rate of reaction is equal in both directions.
- ☐ **D** Equilibrium reactions only occur in liquids.

Q1 When ammonium chloride is heated, it breaks down into ammonia and hydrogen chloride. The reaction is reversible.

$$NH_4Cl_{(s)} \rightleftharpoons NH_{3(g)} + HCl_{(g)}$$

a) The forward reaction is endothermic.

i) Will the backwards reaction be endothermic or exothermic? Tick **one** box.

☐ **A** Endothermic ☐ **B** Exothermic

ii) Which direction will the reaction go in if the reaction vessel is cooled?

....

b) How does the amount of energy transferred in the forward reaction compare to the amount of energy transferred in the backward reaction?

....

c) This reaction can reach equilibrium.
What conditions are needed for equilibrium to be reached? Tick **one** box.

- ☐ **A** The reaction is carried out at a high temperature.
- ☐ **B** The reaction is carried out at high pressure.
- ☐ **C** The reaction vessel is sealed so that the products cannot escape.
- ☐ **D** The backward reaction happens at a faster rate than the forward reaction.

Backwards — it's the new forwards...

Who'd have thought that some reactions can go in two directions... Look out for the little reversible arrow symbol. The direction that a reaction is going in depends on the reaction conditions — so things like temperature and pressure.

Hydrocarbons and Fractional Distillation

Hydrocarbons come in many different sizes, which can be separated using fractional distillation.

Warm-Up

Hydrocarbons are found in <u>crude oil</u> and are made up of only <u>carbon</u> and <u>hydrogen</u> atoms. The <u>properties</u> of hydrocarbons change as their <u>carbon chains</u> get longer.

Circle the correct words to complete the following sentences.

Crude oil is formed mainly of ancient **plankton / coral** that was buried in mud.

Crude oil is a **mixture / compound** of different molecules.

Most of the compounds in crude oil are **carbohydrate / hydrocarbon** molecules.

Crude oil is a **renewable / finite** resource.

Q1 Figure 1 shows a fractionating column used to separate hydrocarbon fractions. The labels for the diagram are missing.

Figure 1

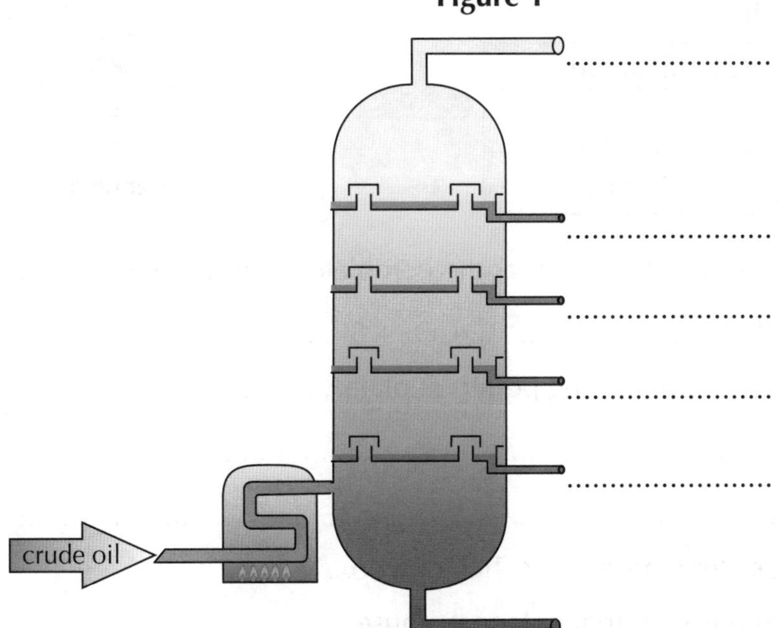

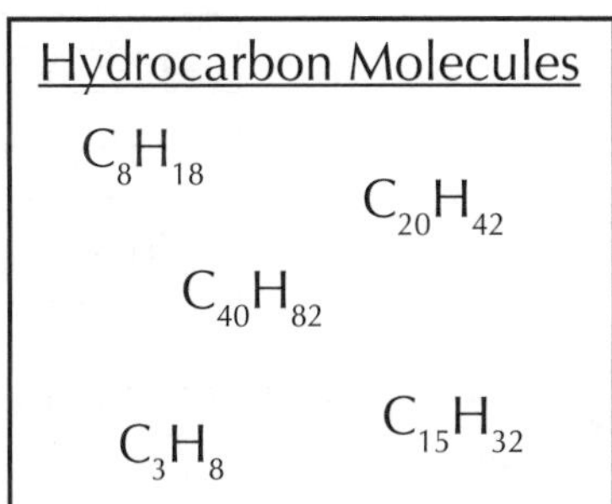

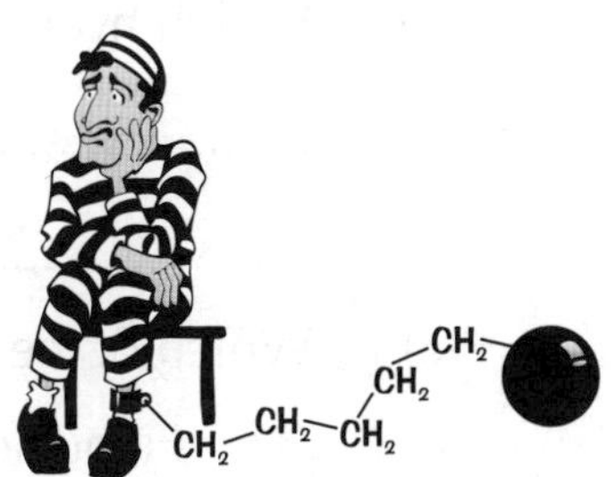

a) Label the fractionating column in **Figure 1** to show which hydrocarbon drains off at each height. Use the hydrocarbons in the box.

b) Which of the following statements is true? Tick **one** box.

- ☐ **A** The top of the fractionating column is the hottest part.
- ☐ **B** The top of the fractionating column is the coolest part.
- ☐ **C** The fractionating column is the same temperature all the way up.

Q2 **Table 1** gives the name, chemical formula and structure of some alkanes. Complete the table by filling in the gaps.

Table 1

Name	Chemical formula	Structure
ethane	....	
....	C_3H_8	
....	....	H H H H | | | | H–C–C–C–C–H | | | | H H H H

Q3 Heptane and triacontane are alkanes.
Table 2 shows the chemical formulas of these two molecules.

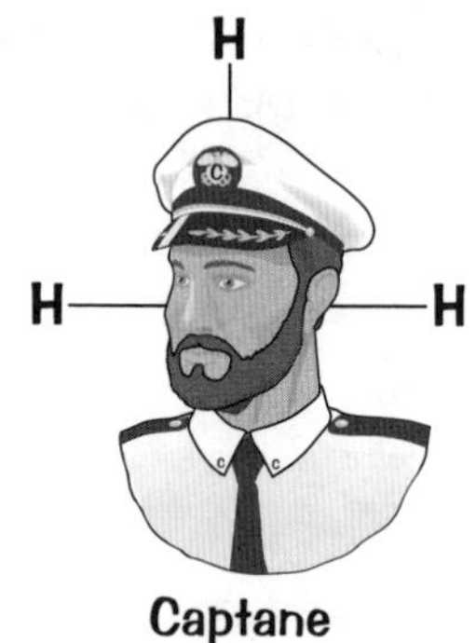

Captane

Table 2

Hydrocarbon	Chemical formula
Heptane	C_7H_{16}
Triacontane	$C_{30}H_{62}$

a) Complete the balanced symbol equation for the complete combustion of heptane.

$$C_7H_{16} + \text{........ } O_2 \rightarrow \text{ .... } CO_2 + 8 \text{ ....}$$

b) Which hydrocarbon has a lower boiling point? Explain your answer.

....

....

c) Which of the following statements is true? Tick **one** box.

☐ Heptane is less viscous than triacontane because heptane has a shorter carbon chain.

☐ Heptane is more viscous than triacontane because triacontane has a longer carbon chain.

d) Which of the two hydrocarbons would be least flammable?

....

Fraction-hating columns — why crude oil can't do maths...

Fractional distillation is an early stage in the processing of crude oil. It separates hydrocarbons from each other according to their chain length. These fractions are then processed further into different materials that are useful to us.

Uses and Cracking of Crude Oil

Crude oil is vital for many things that we rely on in modern life — so let's crack on with it...

Warm-Up

Cracking is a process used to produce more useful molecules from fractions of crude oil.

Tick the boxes to finish the following sentences.

There is a higher demand for...

...short-chain fractions of crude oil. ☐ ...long-chain fractions of crude oil. ☐

Cracking involves heating hydrocarbons to...

...form long-chain hydrocarbons from shorter ones. ☐

...break long-chain hydrocarbons into shorter ones. ☐

Q1 **Figure 1** is a diagram of octane being cracked into hexane and ethene.

Figure 1

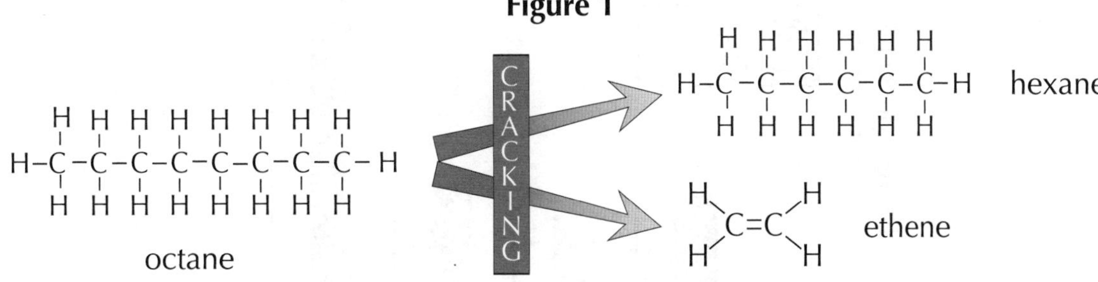

a) Write out the symbol equation for this process.

.... → .... + ...

b) In another reaction, decane ($C_{10}H_{22}$) is cracked to form hexane (C_6H_{14}) and ethene. Complete the balanced symbol equation for this reaction.

$$C_{10}H_{22} \rightarrow C_6H_{14} + \text{..........} \ C_2H_4$$

Q2 Use the words below to fill in the gaps in the paragraph.

alkenes **long** **shorter** **crude oil** **polymers**

Alkanes from .... can be used to make solvents and lubricants.

.... chain alkanes can be cracked to produce alkanes with

.... chains. These make better fuels. Cracking also produces

.... These can be used as the starting material to make ....

Q3 The apparatus in **Figure 2** shows the catalytic cracking of a hydrocarbon.

Figure 2

powdered aluminium oxide

mineral wool soaked in kerosene

gas jar filled with bromine water

boiling tube

Bunsen burner

a) What is the role of aluminium oxide?

....

....

b) The bromine water in the gas jar is bright orange at the start of the experiment. It is colourless at the end. Explain this colour change.

....

....

Q4 Horatio owns a crude oil refinery. He records the amount of each fraction that's present in a sample of crude oil. He then compares it against how much of each fraction his customers want. The results are shown in **Figure 3**.

Figure 3

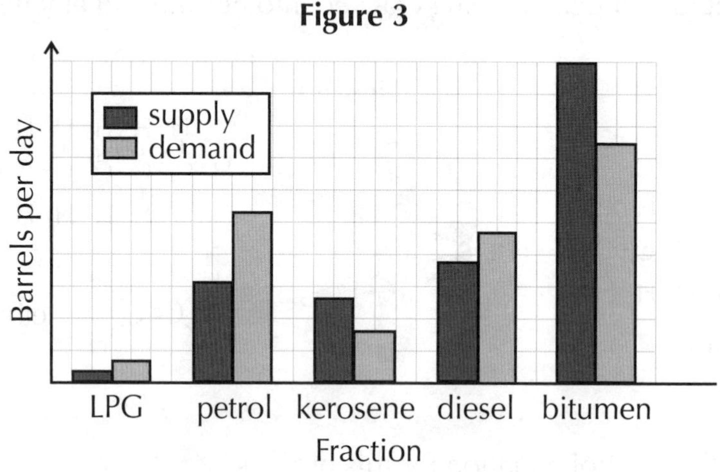

a) The following passage explains why Horatio cracks bitumen to make petrol. Complete the passage by circling the correct words in bold.

The hydrocarbons found in petrol are **shorter / longer** than those in bitumen. This means they are more useful as fuels. **More / less** bitumen is made than is needed. Cracking bitumen **increases / decreases** the **supply of / demand for** hydrocarbons found in petrol. This means less bitumen is wasted.

b) Apart from bitumen, which fraction would be most suitable for cracking to make petrol?

....

You made it through these questions — cracking effort...

Our world would be very different if it wasn't for the discovery of crude oil. It provides the starting materials for loads of different chemicals we use every day, including drugs, fuels, lubricants and polymers. Imagine life without all of that...

Purity and Formulations

It's time to formulate an action plan to tackle this page. Get ready for pure excitement...

Warm-Up

A chemically pure substance is something which contains a single compound or element. The purity of a known substance can be tested by measuring its melting or boiling point. The measured value can then be compared to the known value for the pure substance.

Give two ways that impurities in a sample may affect the melting point of the sample.

1. ....

2. ....

Formulations are mixtures made up of exact amounts of different components. They are made for a specific purpose.

Why do formulations need precise amounts of each component?

....

Q1 A lip balm company makes different lip balms using mixtures of beeswax, coconut oil and peppermint oil. The beeswax is used to give the lip balm structure. The coconut oil is used to soften the lip balm. The peppermint oil is used to add flavour.

The composition of three lip balms, **A**, **B** and **C**, are given in **Table 1**.

Table 1

	Composition (g)		
Component	**A**	**B**	**C**
beeswax	3.5	4	3
coconut oil	3.5	3	4
peppermint oil	0.1	0.1	0.05

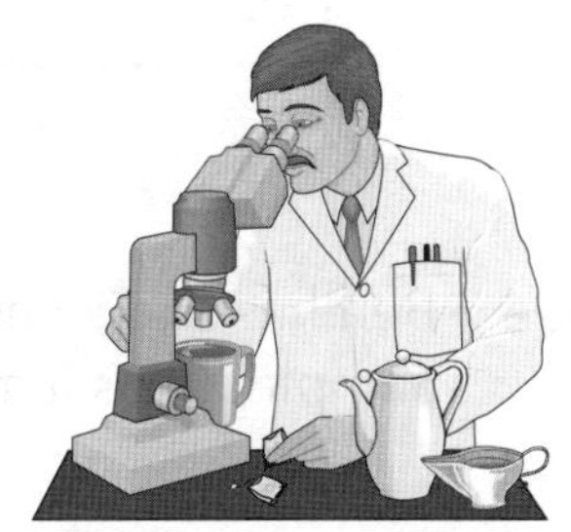

a) Look at the information in **Table 1**. Which of the following are reasons why the lip balms are formulations? Tick **two** boxes.

- [ ] **A** The lip balms contain more than one component.
- [ ] **B** The lip balms are made by mixing measured quantities of the components.
- [ ] **C** The quantities are measured in grams.
- [ ] **D** The quantities are chosen to give the lip balms certain properties.

b) Suggest which lip balm, **A**, **B** or **C**, is the softest. Explain your answer.

....

Q2 Circle each type of formulation in the list below.

shampoo | brass alloy | water | green paint

iron | tin | ammonia | carbon dioxide

steel alloy | wood | bronze alloy | cherryade

Q3 The melting ranges of different samples of glucose were measured.
The results are given in **Table 2**.

Table 2

Sample	A	B	C	D
Melting range (°C)	134-143	136-143	142-144	132-142

a) i) Which of the samples, **A-D**, starts melting at the lowest temperature?

....

ii) Put the samples, **A-D** in order from most impure to least impure.

Most Impure .... **Least Impure**

b) i) Suggest which sample, **A-D**, will have the lowest boiling point.

....

ii) Give a reason for your answer.

....

....

Q4 Ben has two samples of copper, **X** and **Y**. He measures the melting point of each sample.
Sample **X** has a melting point of 1085 °C and sample **Y** melts over the range 910–942 °C.

a) Compare the purity of the two samples.

....

....

....

b) Suggest which of the samples, **X** or **Y**, would have the highest boiling point.

....

55 g of sugar, 125 g of butter and 180 g of flour — best revision formulation...

Remembering how impurities affect substances can be a little difficult. Impurities increase the boiling point, but decrease the melting point, as well as increasing the range over which melting or boiling happens.

Paper Chromatography

PRACTICAL

Paper chromatography is a fairly quick and easy way to find out about certain mixtures.

Warm-Up

Chromatography is used to separate different substances in a mixture.
There are different types of chromatography, including paper chromatography.
All types of chromatography involve two phases.

What are the two phases of chromatography?

....

The spots of chemicals on a chromatogram can be identified by calculating their R_f values and comparing them to reference data.

The formula for calculating the R_f value of a substance is shown below.
Complete the labels on the diagram of the chromatogram.

$$R_f = \frac{\text{distance travelled by substance}}{\text{distance travelled by solvent}}$$

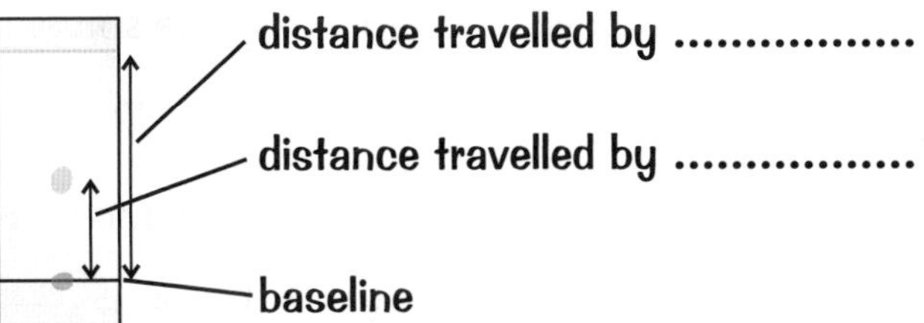

Q1 A food colouring which contained two dyes, **A** and **B**, was analysed using paper chromatography. The chromatogram shown in **Figure 1** was produced.

a) The chromatogram shows two spots. One spot is formed by dye **A** and the other spot is formed by dye **B**.

Using **Figure 1**, calculate the R_f values for dyes **A** and **B**. Give your answers to two significant figures.

Figure 1

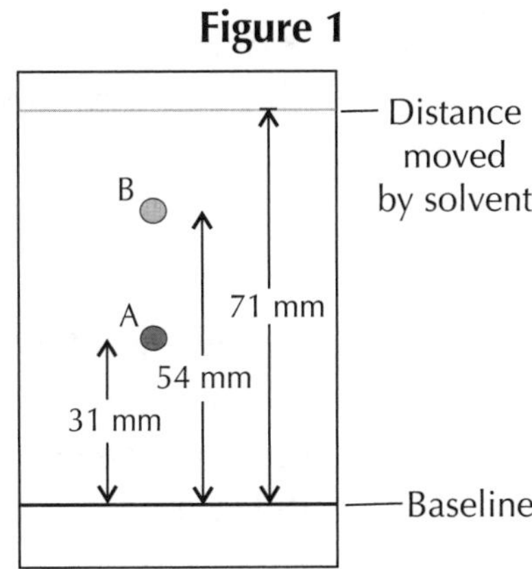

R_f value of **A** = ... R_f value of **B** = ...

b) Dyes **A** and **B** are both attracted to the chromatography paper by a similar amount. Which dye is the most soluble in the solvent?

....

c) Is the food colouring a pure substance? Explain your answer.

....

....

Q2 A forensic scientist used paper chromatography to analyse the ink on a document. The chromatogram she produced is shown in **Figure 2**.

Figure 2

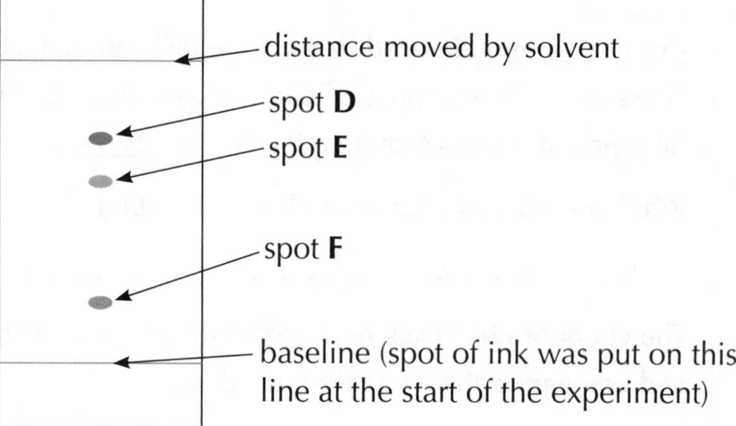

a) i) Use a ruler to measure the distance moved by spot **E** and the distance moved by the solvent.

Distance travelled by spot **E** = ... cm

Distance travelled by the solvent = ... cm

ii) Calculate the R_f value of spot **E**.

R_f value of **E** = ...

b) Would the R_f value of spot **E** be the same or different if the scientist produced another chromatogram using a different solvent? Give a reason for your answer.

....

....

c) The scientist knows that the ink on the document came from one of three printers. She plans to carry out chromatography experiments to separate the ink from each printer. She will then compare the chromatograms produced with the chromatogram in **Figure 2**.

Explain how her chromatograms could be used to identify which of the three printers produced the document.

....

....

....

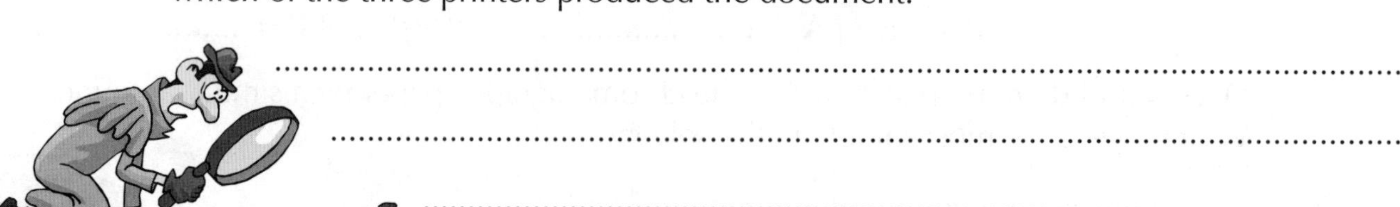

Spending all your money on pens — it's just a stationery phase...

Paper chromatography works because different substances spend different amounts of time in each of the two phases. This means that some substances travel further up the chromatography paper than others, and so have different R_f values.

Tests for Gases

PRACTICAL

This is a small section, but there's still important stuff here. So get ready and let's go...

Warm-Up

A chemical test is a quick way to identify a gas in a lab.
However, there is no single magical chemical test that will tell you what gas you've got.

Each test will only tell you whether or not a certain gas is present. If you really have no clue what the gas is, you have to run the tests one by one until you get lucky.

Describe a chemical test that you could carry out to see if the gas produced in a reaction is carbon dioxide.

....

....

Q1 Nithika carries out a reaction which produces a mystery gas. She collects the gas in a test tube.

a) Nithika puts a glowing splint into the test tube containing the gas. She doesn't observe anything. What does this tell you about the mystery gas?

....

b) Next Nithika puts damp litmus paper in the gas.

i) What gas could she be testing for?

....

ii) Describe the result that would indicate this gas is present.

....

c) Nithika then carries out another test. She listens out for a squeaky pop. Which gas is she testing for?

....

d) Give an example of a safety precaution Nithika should have taken whilst carrying out these tests.

....

....

Required: one non-immature joke about gas...

You only need to know the tests for four gases — so make sure you know them inside out and upside down. If you're having trouble remembering them all, write them out a few times — or even draw a picture for each test.

The Evolution of the Atmosphere

Much like fashion, the atmosphere has changed over time — some changes good, some not so good.

Warm-Up

Earth's atmosphere has changed a lot over time.
There are a number of theories for how it has changed.
It's hard gathering evidence to support these theories. This is because the changes happened over a very long time.

Complete each of the sentences below by ticking the correct box.

In the first billion years of Earth's life, gases in the atmosphere were given out by...

...photosynthesis. ☐ ...volcanoes. ☐ ...charities. ☐

At this time, one of the most abundant gases in the atmosphere was...

...carbon monoxide. ☐ ...carbon dioxide. ☐ ...methane. ☐

This gas was then absorbed by green plants and algae. These organisms produced...

...oxygen. ☐ ...nitrogen. ☐ ...ammonia. ☐

This led to the atmosphere we have today.

Q1 Draw lines to put the statements in the right order on the timeline. One has been done for you.

Carbonate precipitates form in the ocean.

Small amounts of methane and ammonia are released into the atmosphere.

Animals begin to evolve.

The atmosphere is mostly nitrogen and oxygen.

Q2 The pie charts in **Figure 1** show the make-up of gases in the atmosphere during two different time periods.

Figure 1

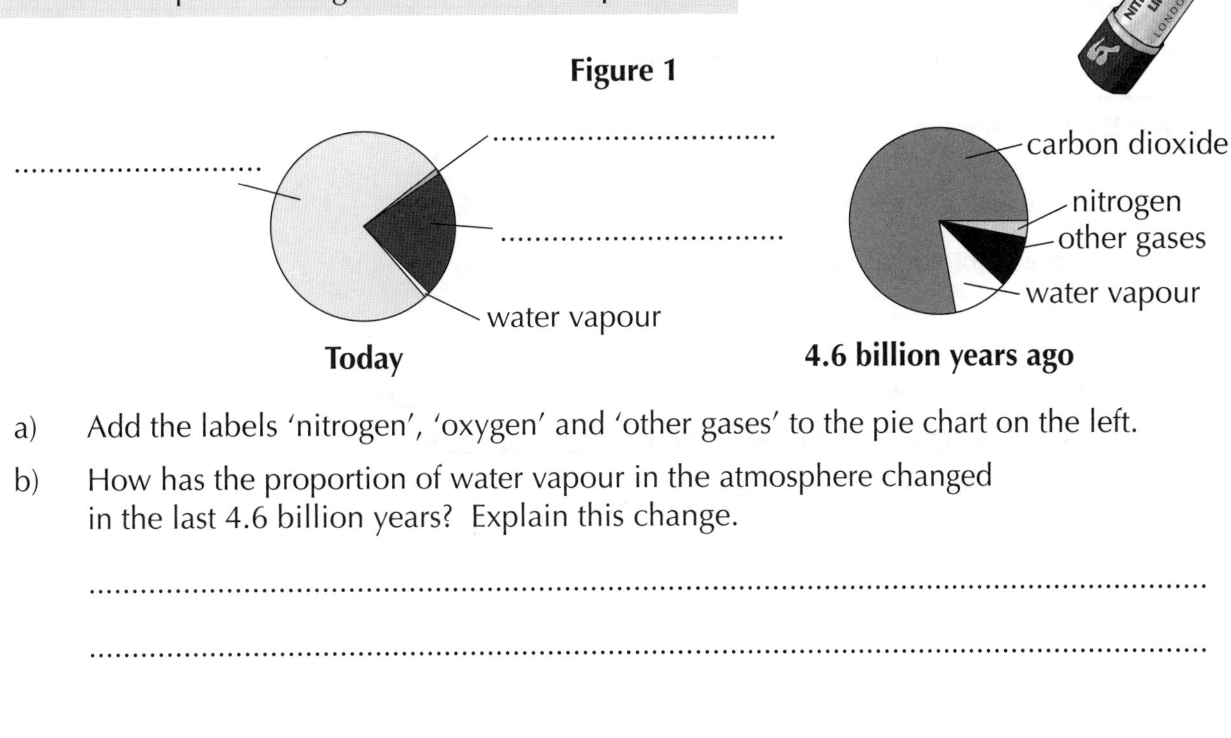

a) Add the labels 'nitrogen', 'oxygen' and 'other gases' to the pie chart on the left.

b) How has the proportion of water vapour in the atmosphere changed in the last 4.6 billion years? Explain this change.

....

....

Q3 The passage below describes how fossil fuels are formed.

a) Complete the passage using the words in the box. You don't need to use every word.

hundreds	marine organisms	sedimentary	carbonate	millions
plant	crude oil	fossil	algae	plankton

Organisms in the sea fall to the seabed when they die. They get squashed down by other dying organisms over .... of years. This process forms .... rocks, such as limestone. Limestone is formed from the shells of dead .... which settle on the seabed. Natural gas and .... are formed from the remains of organisms called ....

b) Why can't coal be found on planets which have never supported life?

....

....

....

Oceans are trendsetters — they were water vapour before it got cool...

You should be able to say how the atmosphere evolved into today's wonderfully life-giving sphere of, erm... gas. It's worth knowing that fossil fuels take millions of years to form and sadly our supply of them won't last forever.

Greenhouse Gases and Climate Change

It's not that relevant, but you may wish to know that I once grew a massive tomato in my greenhouse.

Warm-Up

Greenhouse gases include carbon dioxide and methane.

How do greenhouse gases help to support life on Earth?

....

The levels of greenhouse gases in the atmosphere have increased due to human activities. Most scientists agree that this will lead to climate change.

Give two potential effects of climate change.

1. ....

2. ....

Why is it hard to make a model of the Earth's climate that isn't over-simplified?

....

Q1 Jiang is comparing the effects of nitrogen and carbon dioxide on long wavelength radiation. **Figure 1** shows a diagram of his apparatus.

Figure 1

Long wavelength radiation is thermal (heat) radiation.

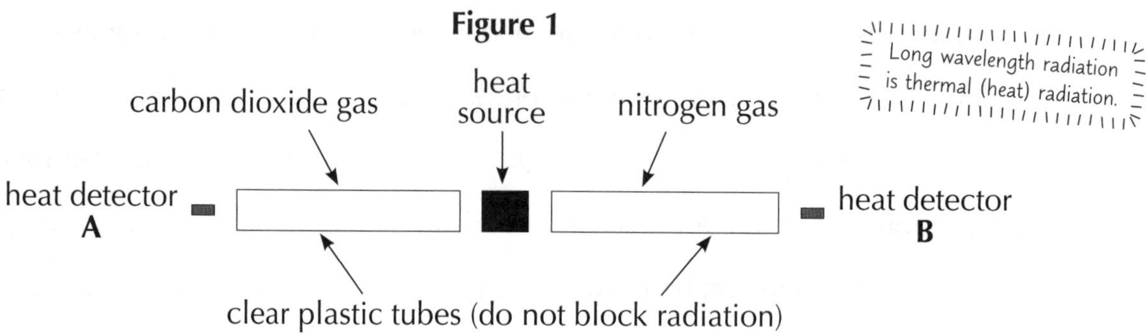

The heat source gives out long wavelength radiation.
This is detected by heat detectors **A** and **B** which are each placed at the end of a plastic tube.
One tube contains nitrogen gas, the other tube contains carbon dioxide gas.

a) Which heat detector, **A** or **B**, will detect more heat from the heat source?

....

Think about which gas will absorb more radiation.

b) Explain your answer.

....

....

....

....

Q2 Answer the questions below using the information shown in **Figure 2** and **Figure 3**.

Figure 2

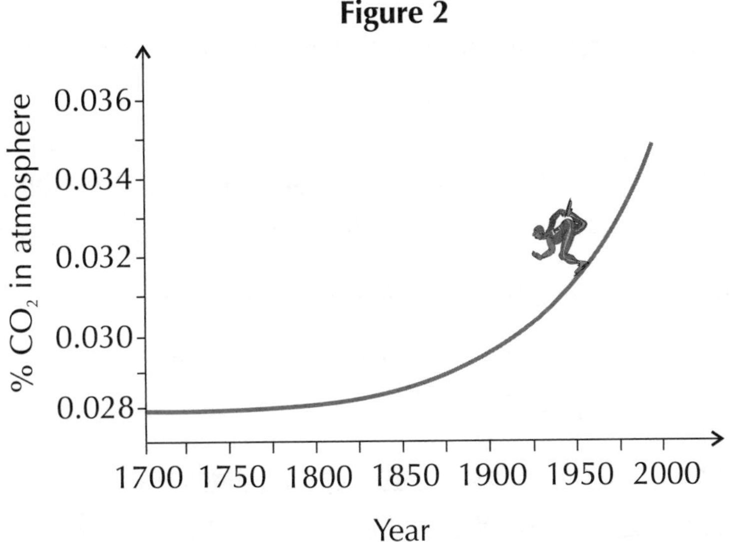

a) Describe the trend shown by **Figure 2**.

....

....

b) Suggest **one** human activity that may have contributed to this trend.

....

c) **Figure 3** shows the average global temperature between 1850 and 2000.

Figure 3

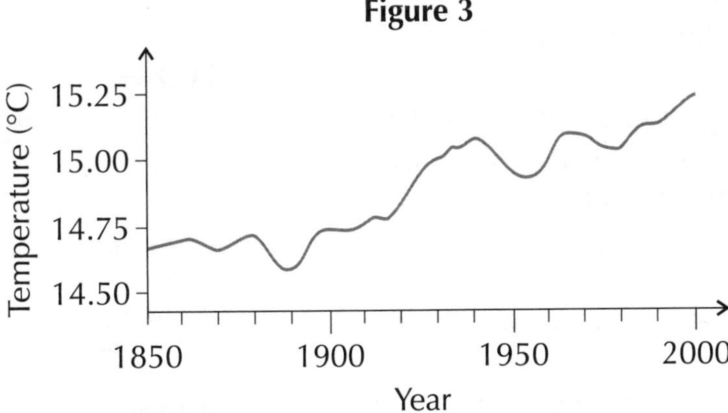

Draw your conclusions from these two graphs only.

Which of the following statements is a valid conclusion that can be drawn from **Figure 2** and **Figure 3**?

☐ **A** The increase in CO_2 levels has caused a rise in global temperature.

☐ **B** CO_2 levels and increasing temperature are positively correlated.

☐ **C** Increasing CO_2 levels are causing climate change.

☐ **D** Increasing global temperatures are causing an increase in CO_2 levels.

Cutting greenhouse gas production — that's emission possible...

Making predictions about the climate can be tricky because there are so many variables that can affect it. Even if there is a correlation between a variable and climate change, it doesn't always mean that one caused the other.

Carbon Footprints and Air Pollution

Keep your mind free from polluting thoughts about dinner tonight, and you'll breeze through this.

Warm-Up

A carbon footprint is a measure of the amount of carbon dioxide, methane and other greenhouse gases released by something over its whole life cycle.

Tick the correct box to show whether the following statement is true or false.

"It is very easy to accurately measure the carbon footprint of a product."

True ☐ **False** ☐

When fuels are burned, they can release carbon particulates and polluting gases. These include sulfur dioxide, oxides of nitrogen and carbon monoxide.

Circle the correct options below to complete the sentences:

Carbon monoxide / **nitrogen dioxide** is a product of incomplete combustion.

Carbon monoxide / **nitrogen dioxide** can cause acid rain.

Global dimming is an effect caused by **carbon particulates** / **sulfur dioxide**.

Q1 **Figure 1** is an extract from Milly's diary.

Figure 1

8.00 am:
Boiled my kettle, which was filled to the top, to make a cup of tea.
Ate a lovely pineapple from Brazil.
10.00 am:
Drove to the travel agent in my sports car and booked a holiday to Hawaii.
11.00 am:
Went shopping for a grass skirt — a must have!
11.30 am:
Got home and put my new skirt on.
I got cold, so I put the gas heating on high. Toasty.
5.00 pm:
Turned all lights on in the house to keep burglars away, then left to catch a plane.

a) Milly's travel agent says that pineapples should be banned in the UK. She argues that transporting pineapples across the world releases a lot of carbon dioxide.

Give **one** reason why people might not agree with banning pineapples in the UK.

....

....

....

Free the pizzas!
BAN PINEAPPLES!

b) Suggest **three** things that Milly could change during her day to reduce her own carbon dioxide emissions.

1. ....

2. ....

3. ....

Q2 Explain why carbon particulates are more likely to form when fuels are burned in engines than when they are burned in the open air.

....

....

....

Q3 A student burns a small sample of a fuel. She collects some of the gas produced in a test tube. The test tube contains a piece of damp litmus paper which turns red. This shows that an acid is present.

Suggest an impurity that might be present in the fuel.

....

Q4 When a power station burns fossil fuels, pollutants including oxides of nitrogen are produced. These pollutants can cause respiratory problems.

a) i) Name **one** other pollutant gas that can cause respiratory problems that may be produced when the power station burns fossil fuels.

....

ii) What causes the production of this gas?

....

....

b) Which of the following statements explains how oxides of nitrogen form at the power station? Tick **one**.

☐ **A** Nitrogen impurities in the fossil fuels react with oxygen in the air when the fuels are burned.

☐ **B** The burning of fossil fuels produces enough heat for nitrogen and oxygen in the air to react with each other.

☐ **C** Nitrogen oxides are formed by the incomplete combustion of the fossil fuels.

☐ **D** Oxygen impurities in fossil fuels react with nitrogen in the air when these fuels are burned.

c) Explain how oxides of nitrogen produced by the power station could lead to acid rain.

....

....

Rain, rain, go away — you've ruined all my gnomes...

Reducing air pollution and cutting our carbon footprints are quite hot topics these days. Companies and governments are doing more to help the situation, so my new gnomes might not be ruined quite as fast. Thank goodness for that.

Finite and Renewable Resources

Earth's resources will run out someday — just like my inspiration for witty page introductions...

Warm-Up

Natural resources are resources which form without input from humans. They can come from the earth, sea or air. The rate at which a natural resource forms affects whether it can be considered as finite or renewable.

Natural resources can be replaced or improved by man-made processes. Agriculture (farming) can be used to help make natural resources meet our needs better.

Complete the following table by putting the resources below in the correct columns.

food uranium

rubber aluminium

fresh water petrol

Finite resources	Renewable resources

Q1 Which of the following statements is true? Tick **one**.

☐ **A** Timber is a natural, non-renewable resource. It is used up more quickly than it can renew itself.

☐ **B** Timber is a natural, renewable resource. It can renew itself more quickly than it is used up.

☐ **C** Timber is a natural, renewable resource. It is used up more quickly than it can renew itself.

☐ **D** Timber is a natural, non-renewable resource. It can renew itself more quickly than it is used up.

Q2 Diesel is a man-made, finite resource. It is used as a fuel.

Diesel is made using crude oil. Explain why diesel is considered to be a finite resource.

....

....

....

Q3 Leslie lives on Planet Mollim. Leslie has discovered three resources which could be used as fuels to power Planet Mollim. **Table 1** shows the energy stored in these resources.

Table 1

Resource	Energy Stored (MJ/m³)
Nababa Fruit Skins	5.0×10^5
Angry Hair Plants	5.0×10^2
Flapadron Tears	2.7×10^7

a) Leslie wants to compare the amounts of energy stored in Nababa Fruit Skins and Angry Hair Plants.

i) Give the energy stored in Nababa Fruit Skins in decimal form.

.... MJ/m³

ii) Give the energy stored in Angry Hair Plains in decimal form.

.... MJ/m³

iii) Calculate the difference between the energy stored in Nababa Fruit Skins and Angry Hair Plants. Give your answer in decimal form.

Difference in energy stored = ... MJ/m³

Leslie has predicted the amount of each resource that will be needed to provide enough energy to power Planet Mollim. **Figure 1** shows Leslie's predictions compared with the amount of each resource formed per decade.

Figure 1

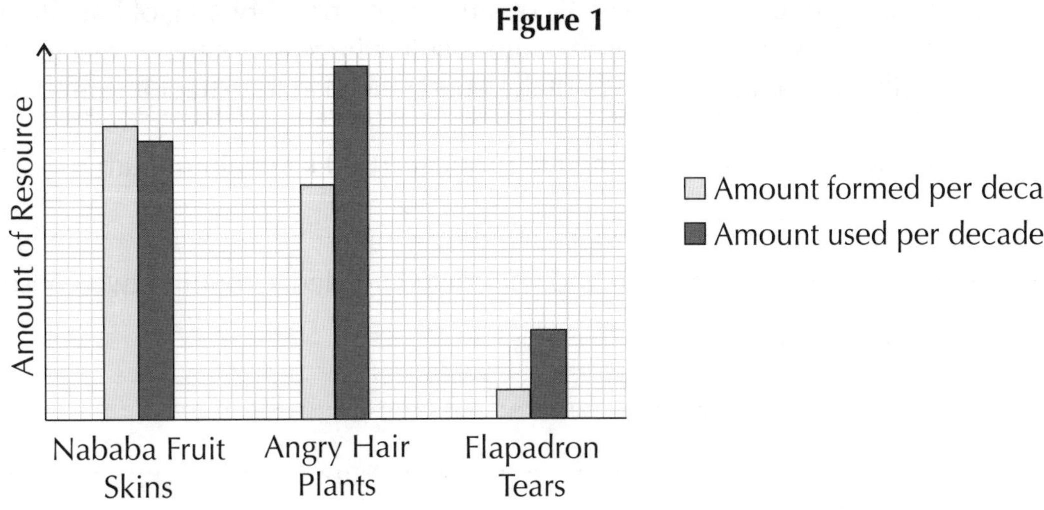

b) Which fuel is most likely to be considered a renewable fuel? Explain your answer.

....

....

CGP — renewing interest in science since the nineties...

It's important that you understand that some resources are renewable, and that others will run out. Lots of companies are now trying to find ways to limit their use of finite resources and develop their businesses in a more sustainable way.

Reuse and Recycling

When you've finished with this book, you could reuse it as a coaster. Or a sledge. Or a frisbee...

Warm-Up

It's important that we try to use resources in a sustainable way.
This means we must think about the needs of people in the future as well as our own needs.
One way of using resources sustainably is to reuse and recycle them.
This helps to stop them from running out.

Reusing glass can help sustainability by reducing the amount of waste produced when glass is thrown away.

Give one other way in which reusing glass helps sustainability.

....

Some forms of glass can be reused without reshaping.
Other forms of glass need to be recycled instead.

How are glass bottles recycled to make new glass products?

....

Q1 Below is some information about aluminium, a widely used metal.

> Bauxite (aluminium ore) gives 1 kg of aluminium for every 4 kg of bauxite mined.
> Extracting aluminium from bauxite requires huge quantities of electricity.
> Recycling aluminium uses 5% of the energy needed to extract it from bauxite.

a) How much ore has to be mined to produce 1000 kg of aluminium?

....

b) Using the information given and your own knowledge, outline **two** consequences of:

i) Mining the bauxite.

1. ....

2. ....

ii) Not recycling the cans.

1. ....

2. ....

Want to hear a joke about potassium? K...

Cracking jokes like the ones you'll find in this book have to be mined from the Earth's crust, you know. So to save energy and reduce the environmental costs of this book, I thought I'd recycle that beauty from page 110.

Life Cycle Assessments

The total impact of the products we make and use needs to be assessed.

Warm-Up

Before a company can produce a new product, they have to carry out a <u>life cycle assessment</u>.

Complete the passage about life cycle assessments using words from the box below.

stage	environment	assess	sustainable	process	materials	protect	cost

Companies have to .... the impact their processes and products will have on the .... They can use this information to choose a .... that does minimal harm. It also helps them to choose the best .... for the job. They have to look at the impact of each .... of the product's life.

Q1 Which stages of a product's life are being described below? Draw lines to match them up.

A computer being powered by electricity.	Using the product.
Poly(ethene) being made from ethene.	Manufacturing the product.
A lot of plastic bottles being thrown away.	Extracting the raw materials.
Oil being drilled out of the ground.	Manufacturing the material.
Window frames being made from PVC.	Disposing of the product.

Q2 Helen is comparing the life cycle assessment for two different CD racks. One is made from metal and the other is made from plastic.

Which stage of the life cycle assessment would be the same for both of the racks? Tick **one**.

- ☐ **A** Extracting and processing the raw materials.
- ☐ **B** Manufacturing the CD racks.
- ☐ **C** Using the CD racks.
- ☐ **D** Disposing of the CD racks.

Q3 Kat is a cafe owner. She wants to replace the plastic straws she gives to customers with a type of straw that is more environmentally friendly. She carries out life cycle assessments for plastic straws and paper straws. Her findings are shown in **Table 1**.

Table 1

	Plastic straw	Paper straw
Raw materials	Crude oil	Timber
Using the product	• Could be washed and reused, but typically are not.	• Can only be used once.
Product disposal	• Too small for most recycling systems, so usually sent to landfill, where they do not biodegrade. • Often end up in rivers and oceans.	• Can be recycled. • Break down in landfill within 50 days.

a) Give **two** reasons why Kat may decide that paper straws are a more environmentally friendly choice. Use the information given in **Table 1**.

1. ....

....

....

2. ....

....

....

b) Kat sees a new biodegradeable plastic straw for sale.

The life cycle assessment for the new straw claims that they are more environmentally friendly than traditional plastic straws because they break down in compost within three months. Kat finds out that this is only true in special industrial compost facilities.

Which of the following is a problem with the life cycle assessment of the new straws? Tick **one** box.

☐ **A** It only includes information that supports claims that the straws are more environmentally friendly than traditional plastic straws, so it is biased.

☐ **B** The straws are made using a different type of plastic, so they can't be compared.

☐ **C** Traditional plastic straws do not biodegrade, and so the two types of straw should not be compared.

All but one of my straws were stolen. Well, that was the last straw...

Chemists have an important role to play in developing materials and processes that minimise harm to the environment. Being able to interpret life cycle assessments will help you understand the bigger picture, so don't skip these questions.

Treating Water

I'd say it's you that's in for the treat with these next two pages...

Warm-Up

The water we use can come from a range of environmental sources, depending on local conditions. Waste water comes from domestic, agricultural and industrial sources. It needs to be treated before it can be returned to rivers and lakes.

Water is treated to make it potable. What is meant by potable water?

....

Which of these two statements is true? Tick one box.

- ☐ **A** Pure water contains dissolved substances but potable water doesn't.
- ☐ **B** Pure water does not contain dissolved substances but potable water sometimes does.

PRACTICAL

Q1 Eva's teacher gives her three samples of water, **A**, **B** and **C**. One of the samples is sea water, one is rainwater and one is tap water. Eva needs to identify each of the three samples.

a) Eva adds universal indicator to a small portion of each sample to measure the pH. Suggest why Eva did not add universal indicator to the whole of each sample.

....

....

....

b) Eva distils 100 cm^3 of each sample. She uses a mass balance to measure the mass of the salt crystals left after all of the water has evaporated. She records her readings in **Table 1**.

Table 1

Sample	Mass of salt crystals (g)
A	3.57
B	0.01
C	0.00

Which of the samples, **A**, **B**, or **C**, is the sea water? Explain your answer.

....

....

....

Q2 **Figure 1** gives some information about sources of water near to the town of Grizeton.

Figure 1

- Grizeton is close to the coast, providing easy access to sea water.
- There are lakes near to Grizeton. However, the warm climate means the water level drops significantly in summer.
- Large amounts of groundwater exist underground at Grizeton.

a) Some of the potable water supplied to Grizeton comes from treating fresh water. Explain why it is unlikely that Grizeton will obtain fresh water through the desalination of sea water.

....

....

....

....

b) i) Would the lakes or the groundwater would be the best source for fresh water?

....

ii) Explain your answer.

....

....

c) Conisbeck is a lake near Grizeton. Describe the steps that could be used to produce potable water from Conisbeck lake water.

....

....

....

....

d) Grizeton also uses a sewage treatment plant to treat waste water from the town. Which of the following is **not** a stage in the process of treating waste water? Tick **one** box.

☐ **A** The waste water is screened to remove any large bits of material and grit.

☐ **B** The waste water goes through sedimentation tanks.

☐ **C** The waste water is heated so that it evaporates.

☐ **D** Effluent is removed and is treated by biological aerobic digestion.

A teabag, milk and two sugars — that's how I treat my water...

There are loads of different stages involved in water treatment, and you need to be able to remember the lot of them (you lucky thing...). Try drawing out some flowcharts to check you know the order things happen in.

Energy Stores, Systems and Conservation of Energy

Time to get some work done by transferring energy to your pen's kinetic energy store...

Warm-Up

The energy of an object is always stored in one or more of the object's energy stores. A system is an object or a group of objects. When a system changes, energy is transferred.

Energy can be transferred mechanically, electrically, or by heating.

Draw lines to match the type of energy transfer below to its description.

mechanical	the energy transfer when a moving charge does work
electrical	when energy is transferred from a hotter object to a colder object
by heating	the energy transfer when a force does work on an object

Q1 **Table 1** shows some situations which involve an energy transfer. Using options from the box below, state the energy store that energy is transferred to. Each option may only be used once.

Table 1

Situation	Energy is transferred to the...
An apple falling from a tree.	.... energy store.
A pan of soup being heated.	.... energy store.
A hair tie being stretched.	.... energy store.
A battery being charged.	.... energy store.
A ball travelling upwards.	.... energy store.

elastic potential	thermal	kinetic	chemical	gravitational potential

Q2 An electric heater is connected to the mains electricity supply. Which sentence correctly describes the energy transfer between the mains electricity supply and the heater?

- [ ] **A** Energy is transferred electrically to the kinetic energy store of the heater.
- [ ] **B** Energy is transferred by heating to the kinetic energy store of the heater.
- [ ] **C** Energy is transferred electrically to the thermal energy store of the heater.
- [ ] **D** Energy is transferred by radiation to the thermal energy store of the heater.

Q3 A car is travelling over flat, horizontal ground. The car has 41 000 J of energy in its kinetic energy store. The driver applies the brakes to slow down, causing the brakes of the car to heat up. While braking, 29 000 J of energy is transferred to the thermal energy stores of the car and its surroundings.

Assume the car and its surroundings are a closed system.
How much energy is in the car's kinetic energy store after braking?
Circle the correct answer.

12 000 J   29 000 J   41 000 J   70 000 J

Q4 A weightlifter is holding a set of weights still above her head.

a) Describe the energy transfers involved when the weightlifter raises the weights.

b) The weightlifter drops the weights.
Describe the energy transfer that will take place as they fall towards the floor.

Gift cards for energy stores — that idea's got potential...

You may need to describe the energy transfers occurring in a system, so make sure you can recognise them. For example, if an object is changing temperature, energy is being transferred to or from its thermal energy store.

Kinetic and Potential Energy Stores

If you can master these pages, you've got the potential to go far. Better get a move on then...

Warm-Up

An object that is moving will always have some energy in its kinetic energy store.

An object that is in a gravitational field has energy in its gravitational potential energy (g.p.e.) store.

An object that has been stretched or squashed has energy in its elastic potential energy store.

State the equation that links the energy in an object's kinetic energy store, its mass, and its speed.

....

State the equation that links the energy in an object's gravitational potential energy store, its mass, its height above the ground, and the gravitational field strength.

....

Q1 A roller coaster cart is stationary at the top of a slope. It then rolls down the slope.

a) Use any of the words or phrases in the box to fill in the gaps in the paragraph below.

increases	half	all
decreases	some	stays the same

As the cart rolls down the slope, the speed of the cart increases. This means that the amount of energy in the cart's kinetic energy store ....
If there are no frictional forces acting on the cart, ... of the energy transferred out of the cart's gravitational potential energy store is transferred to the cart's kinetic energy store as it rolls down the slope.

b) At the bottom of the slope, the cart is travelling at a speed of 12 m/s.
The cart has a mass of 500 kg.
How much energy is in the kinetic energy store of the cart at the bottom of the slope?

energy = ... J

Q2 A dog pulls on its lead, causing it to stretch by 0.0050 m. The lead has a spring constant of 280 000 N/m.

a) Calculate the energy stored in the elastic potential energy store of the dog lead when the dog pulls on it. You can assume the limit of proportionality has not been passed.

energy = ... J

b) The dog sees a squirrel and pulls harder on the lead. This causes the lead to extend further. Describe how this affects the amount of energy stored in the lead's elastic potential energy store.

....

....

Q3 Daichi works at a DIY shop. He has to load some flagstones onto the delivery truck. Each flagstone has a mass of 25 kg.

a) To load the flagstones, Daichi has to lift them 1.2 m above the ground. Calculate the energy transferred to the gravitational potential energy store of one flagstone when it is lifted 1.2 m above the ground. The gravitational field strength is 9.8 N/kg.

energy = ... J

b) Once Daichi has loaded all the flagstones onto the delivery truck, he drives it away. The truck is travelling at 10 m/s, and has 150 000 J of energy in its kinetic energy store. Calculate the total mass of the truck.

mass = ... kg

My local greengrocer keeps moving — it's a kinetic store...

In GCSE Science you don't need to remember the equation for energy in an elastic potential energy store. But you need to remember the equations for energy in kinetic and gravitational potential energy stores. Make sure you know them.

Specific Heat Capacity

Whenever there's an increase in temperature, there has to be a transfer of energy. How *much* energy depends on the specific heat capacities of any substances involved...

Warm-Up

Different substances have different specific heat capacities.
Use the correct options from the box below to complete the definition of specific heat capacity.

temperature	0.1 J	mass	10 °C	energy	1 °C

Specific heat capacity is the amount of .... that must be transferred to 1 kg of a substance to increase its .... by ....

Q1 Max puts a frying pan over a flame on a hob. He puts some oil into the frying pan. Energy is transferred to the oil, causing the oil's temperature to rise.

a) Max wants to increase the temperature of the oil by 44 °C. Put ticks in **Table 1** to show whether the statements for increasing the temperature of the oil are true or false.

Table 1

Statement	True	False
The larger the mass of the oil, the more energy Max needs to transfer to the oil.		
The larger the specific heat capacity of the oil, the more energy Max needs to transfer to the oil.		
The lighter the colour of the oil, the more energy Max needs to transfer to the oil.		

b) When Max heats 0.025 kg of oil in the pan, the oil temperature increases by 44 °C. The oil has a specific heat capacity of 1670 J/kg°C. Calculate the energy transferred to the oil.

energy = ... J

c) Max turns the hob off, and the oil cools down. The oil releases the same amount of energy that was transferred to it when it was heated. By how much will the temperature of the oil have decreased when it has finished cooling down? Tick the correct answer.

☐ 22 °C ☐ 44 °C ☐ 88 °C

PRACTICAL

Q2 Janelle is doing an experiment to investigate the specific heat capacities of two different liquids, A and B.

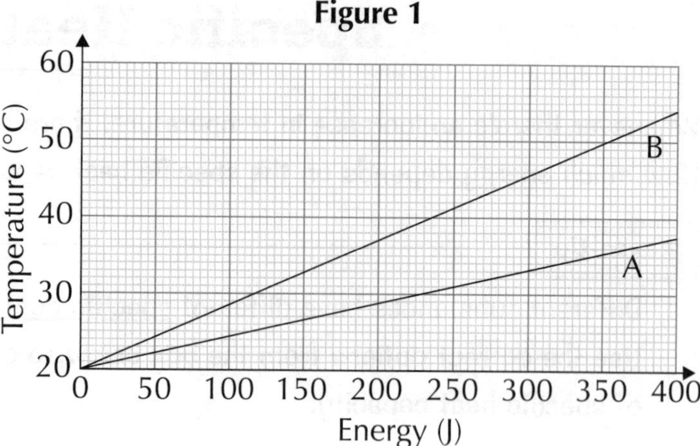

Janelle heats 0.018 kg of each liquid in separate insulated flasks. Each liquid has a total of 400 J of energy transferred to it during the experiment.

She records the temperature of each liquid at certain points throughout her experiment. Her results are shown in **Figure 1**.

a) What was the total temperature change of liquid A in Janelle's experiment?

Be careful — the y-axis doesn't start at 0 °C.

temperature change = ... °C

b) Calculate the specific heat capacity of liquid A.
Give your answer to 2 significant figures.

specific heat capacity = ... J/kg°C

c) Does liquid B have a higher or lower specific heat capacity than liquid A?

Both liquids were given the same amount of energy — think about whether liquid B's temperature increased more or less than liquid A.

....

Q3 Mildred wants to warm up her bed on a cold evening. She decides to use a heated bean bag.

43 000 J of energy is transferred to the beans in the bean bag to increase their temperature by 50 °C. The specific heat capacity of the beans is 1700 J/kg°C. What is the mass of the beans in the bean bag?

mass = ... kg

Fun fact: the specific heat capacity of the Pacific ocean is 3900 J/kg°C (ish)...

Specific heat capacity? Sounds like someone's just thrown three words together and called it physics. Unfortunately you do need to know what it is, how to use it, what its favourite animal is... that sort of thing. Ok, maybe not that last one.

Power and Reducing Energy Transfers

No matter how powerful you are, you can't avoid wasting some energy...

Warm-Up

Energy isn't always transferred to useful energy stores.

Power is a measure of the rate at which energy is transferred.
The unit that power is measured in is the watt (W).

Using any of the words or phrases in the box below, complete the following sentences.

always	chemical	sometimes	useless	gravitational potential

When a system changes, some energy is .... dissipated.

This wasted energy ends up in .... energy stores.

Which of the following is the correct definition of 1 watt? Tick one box.

☐ 1 joule of energy transferred per second

☐ 1 joule of energy transferred per metre

☐ 1 joule of energy transferred per watt

Q1 **Figure 1** shows parts of three different rooms, and some information about their walls.

Figure 1

A

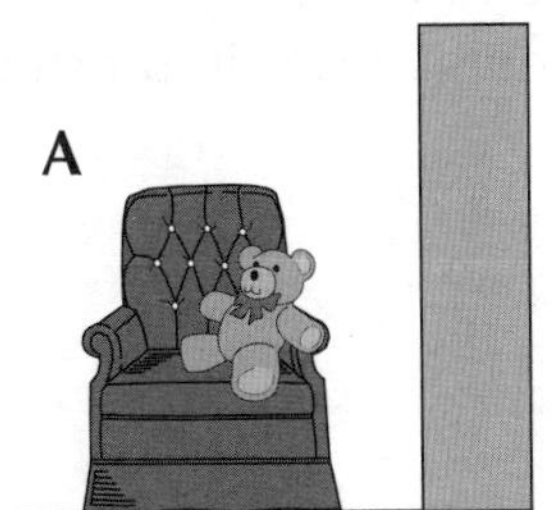

wall thickness = 20 cm
thermal conductivity of wall = 0.15 W/m·K

B

wall thickness = 8 cm
thermal conductivity of wall = 0.40 W/m·K

C

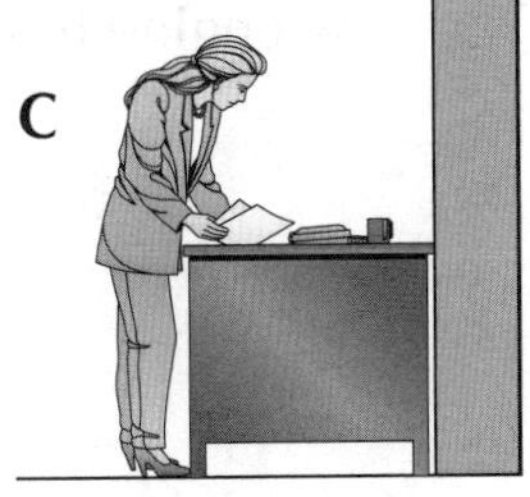

wall thickness = 15 cm
thermal conductivity of wall = 0.35 W/m·K

Don't worry that the units for thermal conductivity look a bit weird. You don't need to understand or know them. Phew.

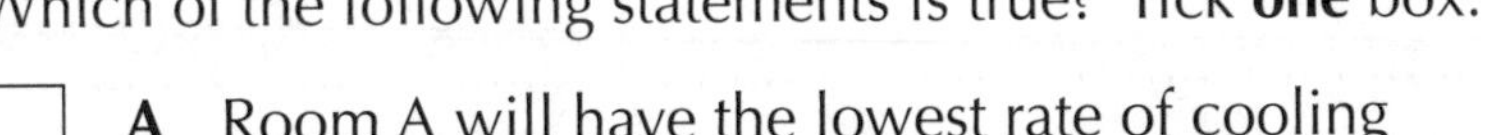

a) Circle the room that will have the highest rate of cooling.

b) Which of the following statements is true? Tick **one** box.

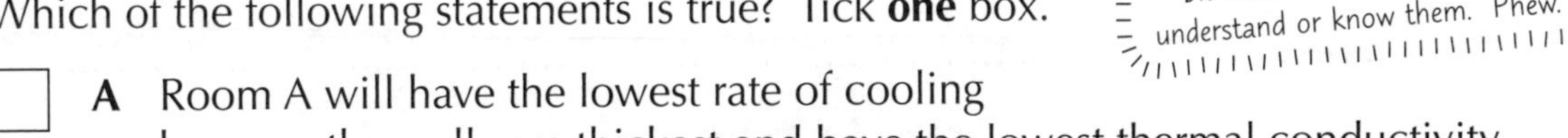

☐ **A** Room A will have the lowest rate of cooling because the walls are thickest and have the lowest thermal conductivity.

☐ **B** Room B will have the lowest rate of cooling because the walls are thinnest and have the lowest thermal conductivity.

☐ **C** Room C will have the lowest rate of cooling because the walls are thickest and have the highest thermal conductivity.

Q2 **Table 1** shows the powers of 3 kettles. Each kettle is used to heat 1 litre of water from 20 °C to 100 °C. You can assume no energy is wasted by the kettles.

Table 1

Kettle	Power (W)
A	900
B	3000
C	2200

a) Which kettle will heat the water to 100 °C fastest?

....

b) Explain your answer.

....

....

....

Q3 A mechanic is fixing a moped.

a) The engine in the moped transfers 12 500 J of energy in 25 s.
Calculate the power of the engine.

power = ... W

b) The mechanic replaces the engine in the moped with a new one.
The new engine has a power of 760 W. How much energy does it transfer in 25 s?

energy = ... J

c) While working on the moped, the mechanic applies a lubricant to the moving parts of the moped. Explain how a lubricant can reduce unwanted energy transfers in the moped.

....

....

....

With great power, comes great practice questions...

Sadly, no radioactive spiders were involved in the making of these questions. But they should still help you get to grips with energy and power. Remember, you can calculate power using energy transfer (or work done) over time.

Efficiency

Test your efficiency by answering these questions without wasting any energy.
That means no checking your phone until you're done.

Warm-Up

Whenever a <u>device</u> is used to <u>transfer energy</u>, some energy will always be <u>wasted</u>.
<u>Efficiency</u> is a <u>measure</u> of the amount of energy that is usefully transferred.

Fill in the box in the equation below, to give the correct equation for efficiency.

$$\text{Efficiency} = \frac{\boxed{}}{\text{total input energy transfer}}$$

Q1 Some information about washing machines during a wash cycle is shown in **Table 1**.
Which of the following washing machines is the most efficient? Tick **one** box.

Table 1

	Washing machine	Input energy (J)	Useful output energy (J)
☐	A	40 000	25 000
☐	B	40 000	28 000
☐	C	40 000	29 500

Q2 5350 J of energy is transferred to an electric whisk. 1220 J of this energy is wasted.

a) What is the useful output energy transferred by the whisk?
Circle the correct answer.

439 J   4130 J   4350 J   6570 J

b) Calculate the efficiency of the whisk as a decimal.

efficiency = ...

c) What is the efficiency of the whisk as a percentage?

efficiency = ... %

Q3 An electric kettle has a power input of 2500 W and an efficiency of 0.72.

a) Which of the following can be used to calculate the useful power output of the kettle? Tick **one** box.

- [ ] **A** useful power output = efficiency ÷ total power input
- [ ] **B** useful power output = efficiency × total power input
- [ ] **C** useful power output = total power input ÷ efficiency
- [ ] **D** useful power output = total power input + efficiency

b) Calculate the useful power output of the kettle.

useful power output = ... W

PRACTICAL

Q4 Viola is investigating how the efficiency of a solar panel changes with temperature. She varies the temperature of the solar panel and measures the power output. During her investigation, the power input to the solar panel is kept at 1675 W. Her results are shown in **Figure 1**.

Figure 1

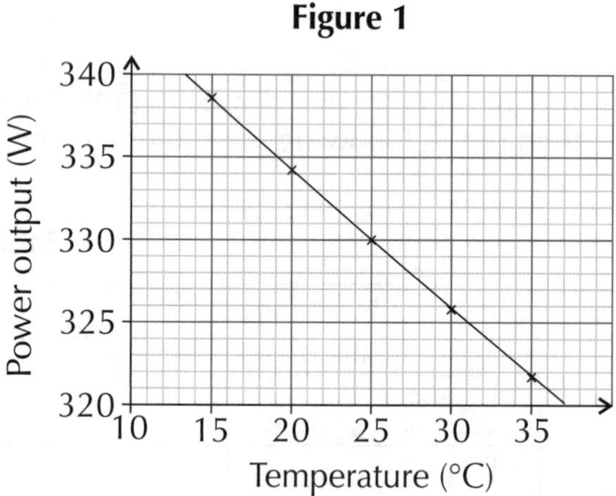

a) Yesterday, the solar panel had a temperature of 25 °C. Calculate what the efficiency of the solar panel was yesterday.

efficiency = ...

b) i) Is the solar panel more efficient at higher or lower temperatures?

....

ii) Explain your answer.

....

....

....

....

A fish ant. A terrifyingly efficient made-up creature...

It's easy to get yourself in a muddle with efficiency calculations, but try not to panic. If you're calculating the efficiency, take your time and make sure you know which is the input energy (or power) and which is the output energy (or power).

Energy Resources and their Uses

So many things in our lives rely on electricity to deliver the energy that makes them work. But where do we get the energy to generate the electricity? Introducing energy resources...

Warm-Up

Energy resources can be renewable or non-renewable. Non-renewable energy resources will eventually run out, but renewable energy resources won't.

Draw a circle around each of the options in the box below that is a fossil fuel.

nuclear fuel   coal   wind   geothermal   natural gas   oil

Draw a circle around each of the options in the box below that is a renewable energy resource.

nuclear fuel   the Sun   the tides   water waves   natural gas   coal

Q1 A small town is located on the top of a hill, far away from the coast. The weather in the town is usually cloudy and overcast, and it regularly experiences fairly strong winds.

Which of the following would provide the most suitable source of electricity in the town? Tick **one** box.

- [ ] **A** tidal barrage
- [ ] **B** wave-powered turbines
- [ ] **C** solar panels
- [ ] **D** wind turbines

Q2 Five students are discussing why they think it's better to use bio-fuels rather than coal to generate electricity.

Put a tick next to each student whose statement is true.

- [ ] **Annie** Bio-fuels don't release any carbon dioxide when they're burnt, but coal does.
- [ ] **Ben** Bio-fuel is a renewable resource, but coal will eventually run out.
- [ ] **Charles** Bio-fuels are safer to use than coal, because coal produces nuclear waste.
- [ ] **Diego** The plants used to create bio-fuels can be replaced with more plants faster than coal can be replaced after it has been extracted.
- [ ] **Esme** Growing plants to make bio-fuels removes some carbon dioxide from the atmosphere, but extracting coal doesn't.

Q3 At a public meeting, the locals are discussing plans to build a hydroelectric dam. **Figure 1** shows some statements they make about hydroelectric dams.

Figure 1

Alisha
We should use hydroelectric power because it doesn't cause any pollution.

Greg
A hydroelectric dam will cause a lot of damage to the countryside.

Eliza
A hydroelectric dam will always give us electricity when it's needed.

Draw a circle around the correct word to say whether you agree or disagree with each statement. Give a reason for each of your answers.

a) i) I **agree** / **disagree** with Alisha because ....

....

....

....

ii) I **agree** / **disagree** with Greg because ....

....

....

....

iii) I **agree** / **disagree** with Eliza because ....

....

....

....

b) The local community have also agreed to use solar panels to provide some of their energy. Suggest why the local community are not using solar panels to provide all of their energy.

....

....

....

Hydroelectric jam — the energy resource you can spread on your toast...

There's a lot to learn about energy resources, but you can make it easier for yourself by tackling each resource separately. For each resource, make sure you learn the main features of that resource, whether it's renewable or not, how reliable it is, and how much of an impact it has on the environment. That should give you plenty of fuel for answering questions.

Trends in Energy Resource Use

They may look similar, but wind turbines are no fidget spinners. Energy trends go on for a while...

Warm-Up

Circle the correct words to complete the sentence below.

Electricity use in the UK has (decreased / increased) recently, as people are more aware about the (environmental / economic) effects of generating electricity from fossil fuels.

Q1 **Figure 1** shows how the amount of power produced in a country by solar panels has changed each year since the year 2000.

Figure 1

Power produced by solar panels in the country

0

00 01 02 03 04 05 06 07 08 09 10 11 12 13 14 15 16 17 18

Year (20XX)

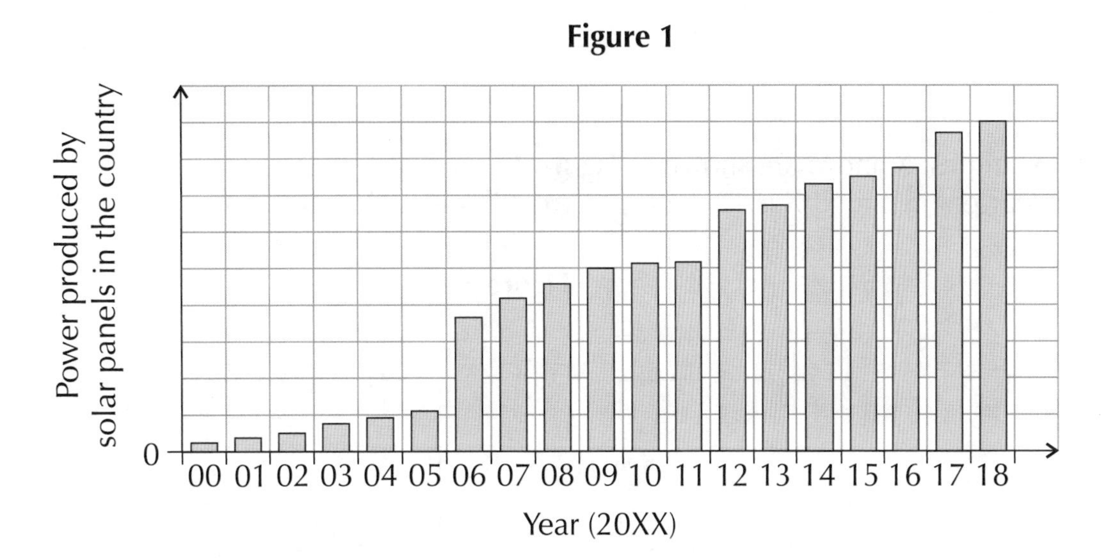

a) The graph in **Figure 1** shows that there was an increase in the number of solar panels being used between 2000 and 2018. Suggest why more people are putting solar panels on their houses each year.

....

....

b) i) At one point, the country's government started paying people if they installed solar panels on their home. In which year is this likely to have happened? Tick **one** box.

☐ **A** 2002 ☐ **B** 2006 ☐ **C** 2010 ☐ **D** 2016

ii) Explain your answer.

....

....

....

I never put enough ketchup on my chips — I always need to re-sauce...

There are many things that can affect how we decide to generate electricity. These include political and business issues, worries about the environment, worries about people's health and happiness, and the cost of using the resource.

Current and Circuit Symbols

Don't get 'current' and 'currant' mixed up — you'd get a shock if you bit into a current teacake.

Warm-Up

Components in a circuit are drawn as <u>circuit symbols</u> in circuit diagrams.

Draw lines to connect each component name to the correct circuit symbol.

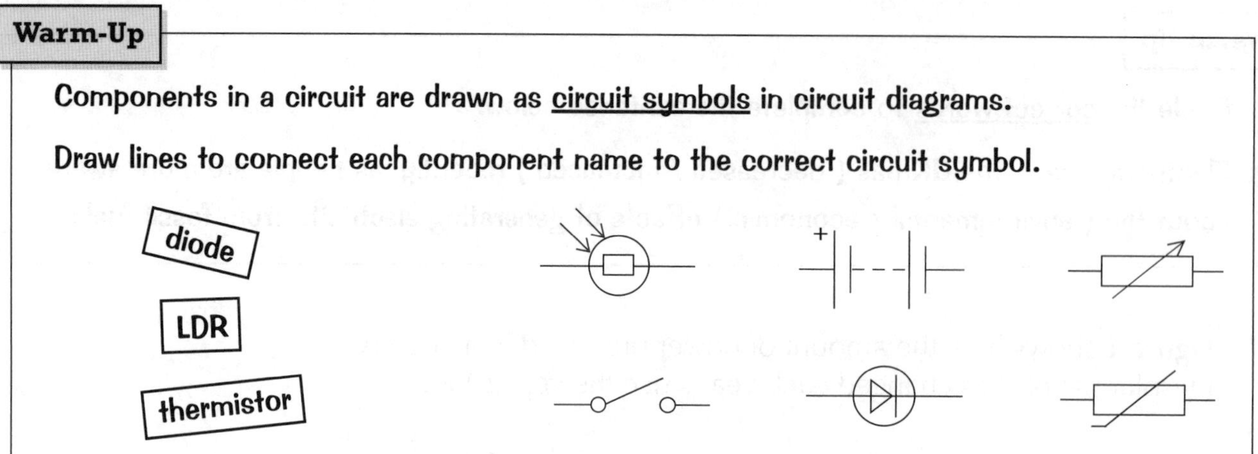

Q1 Two circuits, X and Y, are shown in **Figure 1**.
The current in circuit X is 12 A. The current in circuit Y is 3.0 A.

Figure 1

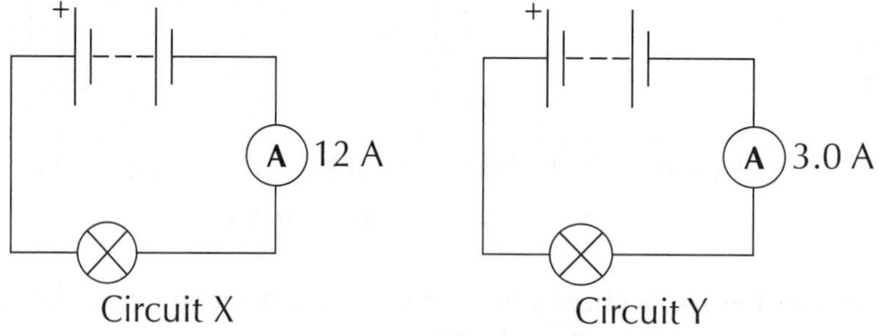

a) Calculate the total charge that flows through the bulb in Circuit X in 5400 s.

charge = ... C

b) Calculate how long it takes the same amount of charge to pass through the bulb in circuit Y.

time = ... s

The new electric-jazz band is nothing without Sir Kit's cymbals...

As well as knowing how to calculate charge flow, you need to remember lots of circuit symbols too.
Make sure you know LEDs from LDRs, resistors from thermistors, and if you're making a tea — sugar from salt.

Resistance and V = IR

Don't resist these questions about resistance. They're irresistible...

Warm-Up

The resistance of a component (R), the potential difference across it (V) and the current through it (I) are all linked by the equation $V = IR$.

Circle the correct word to complete the following sentence:

If the potential difference across a component is kept the same, but its resistance increases, then the current through the component decreases / increases .

Q1 A student makes a circuit that contains a battery and resistor A, shown in **Figure 1**. The potential difference across resistor A is 6.0 V. Resistor A has a resistance of 2.5 Ω.

Figure 1

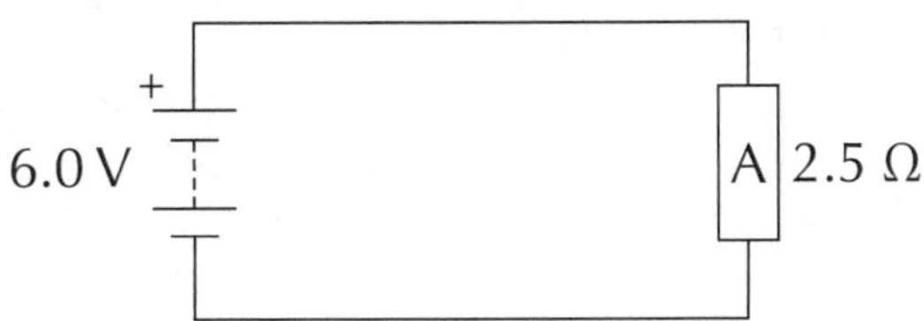

a) i) Circle the equation below that can be used to find the current through resistor A.

$I = V \times R$ $I = V \div R$ $I = R \div V$ $I = R \times V$

ii) Calculate the current through resistor A.

current = ... A

b) The student replaces resistor A with another resistor, resistor B.
The potential difference across resistor B is 6.0 V and the current through it is 0.75 A.

Calculate the resistance of resistor B.

resistance = ... Ω

c) The student replaces the old battery with a different one so the current through resistor B is now 0.5 A. Which of the following is true? Tick **one**.

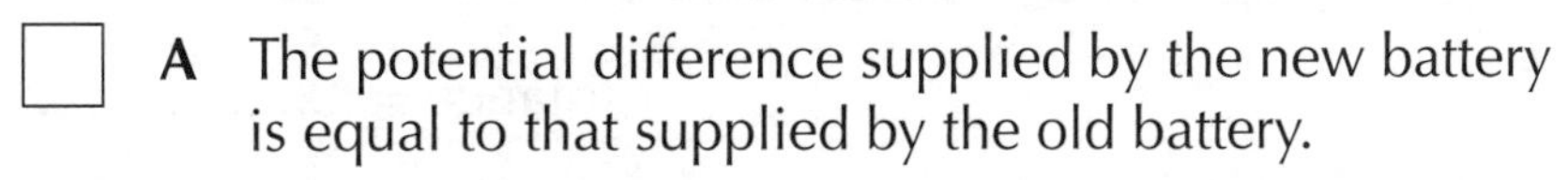

☐ **A** The potential difference supplied by the new battery is equal to that supplied by the old battery.

☐ **B** The potential difference supplied by the new battery is higher than that supplied by the old battery.

☐ **C** The potential difference supplied by the new battery is lower than that supplied by the old battery.

Rest, Sister Bea

PRACTICAL

Q2 Shimla did an experiment to see how the resistance of a wire changes with its length. Shimla set up the circuit shown in **Figure 2**.

Figure 2

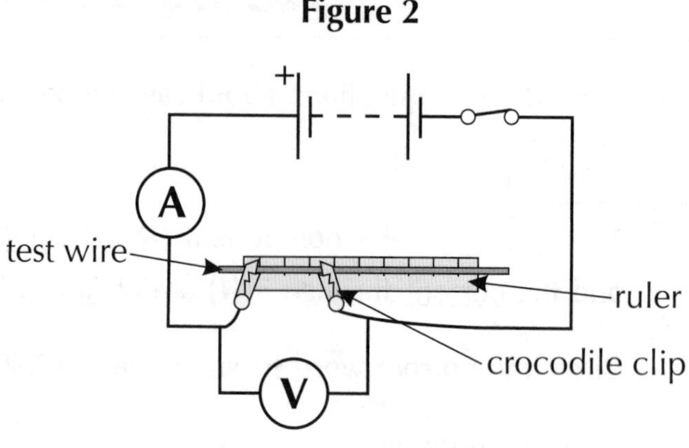

Shimla changed the length of the test wire in the circuit by moving the crocodile clips. For each length of test wire, Shimla recorded the potential difference across the test wire and the current through it.

Shimla then calculated the resistance of each length of test wire. Her results are shown in **Figure 3**.

Figure 3

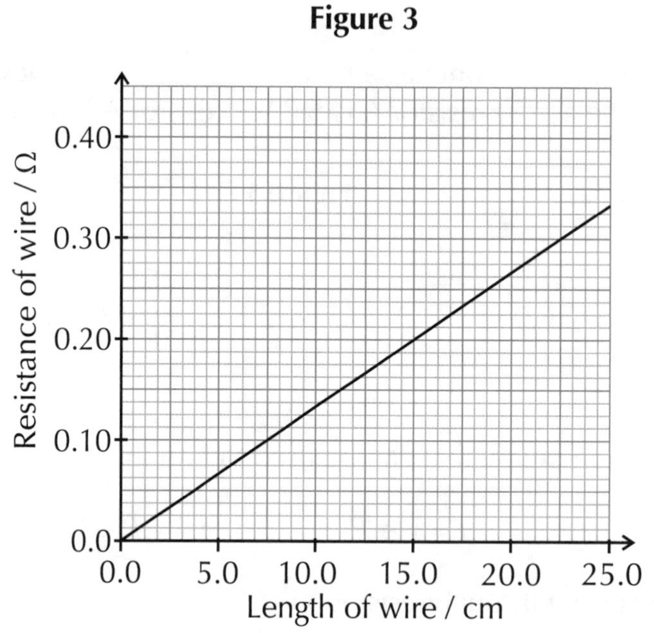

a) Use **Figure 3** to determine the resistance per cm of the test wire. Give your answer in Ω/cm.

Hint: resistance per cm = $\frac{\text{resistance of wire}}{\text{length of wire}}$

resistance per cm = ... Ω/cm

b) Use your answer from a) to estimate the resistance of 60 cm of the test wire.

resistance = ... Ω

If I had a higher resistance to chocolate, I might be able to stick to a diet...

You might see questions about experiments that don't match what you've done in class, e.g. finding the resistance per cm. Don't let this throw you off — think about how to apply what you already know to the new question.

I-V Characteristics and Circuit Devices

Ivy is a climbing plant. It can grow indoors and outdoors and can reach heights up to 30 m. Some bugs feed on leaves and twigs of ivy and... What's that? Oh. *I-V* characteristics. I see...

Warm-Up

The resistance of some components changes as the current through them changes.
The resistance of ohmic conductors doesn't change as the current through them changes.

Use the correct words or phrases from the box to complete the sentences below:

decreases	increases	burglar detectors	thermostats	stays the same

LDRs are components whose resistance depends on light intensity.

As the light intensity increases, the resistance ....

LDRs are useful in things like ....

An *I-V* characteristic is a graph which shows how the current through a component changes with potential difference.

Which one of the following components produces an *I-V* characteristic like the one on the right?

- ☐ **A** filament lamp
- ☐ **B** diode
- ☐ **C** ohmic conductor

Current

Potential difference

Q1 A thermistor is used in a central heating thermostat.

a) Circle the correct word or phrase to complete the following sentence:

> The resistance of a thermistor **increases / decreases / stays the same** as the temperature increases.

When the temperature drops below 20 °C, the thermostat turns the central heating on.
At 20 °C the resistance of the thermistor in the thermostat is 9100 Ω.

b) At one moment, the temperature is 18 °C and the central heating is on.
What could the resistance of the thermistor be at this moment? Tick any that apply.

☐ 4900 Ω ☐ 9000 Ω ☐ 9400 Ω ☐ 12 500 Ω

PRACTICAL

Q2 A student set up a circuit to identify an unknown component, component X. This circuit is shown in **Figure 1**.

Figure 1

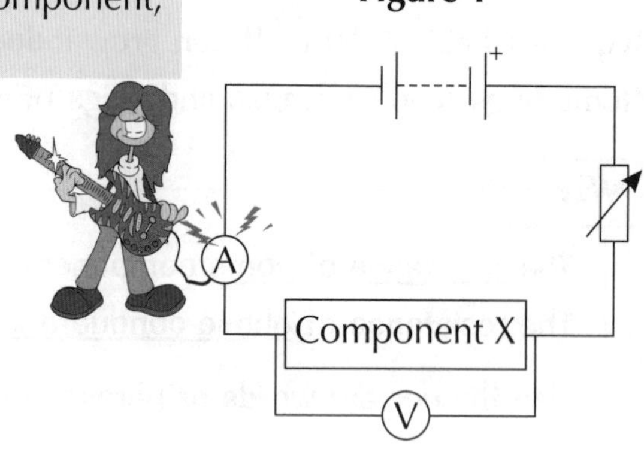

She used the variable resistor to change the potential difference across component X. She recorded several different potential differences and the current flowing through component X at each potential difference.

Table 1 shows her readings of current and potential difference for component X.

Table 1

Potential difference (V)	–4.0	–3.0	–2.0	–1.0	0.0	1.0	2.0	3.0	4.0
Component X current (A)	0.0	0.0	0.0	0.0	0.0	0.20	1.0	2.0	4.5

a) Plot an *I-V* characteristic for component X on **Figure 2**.

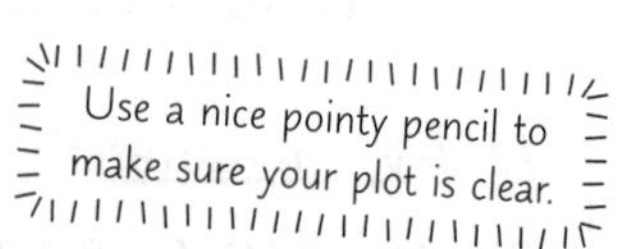

Figure 2

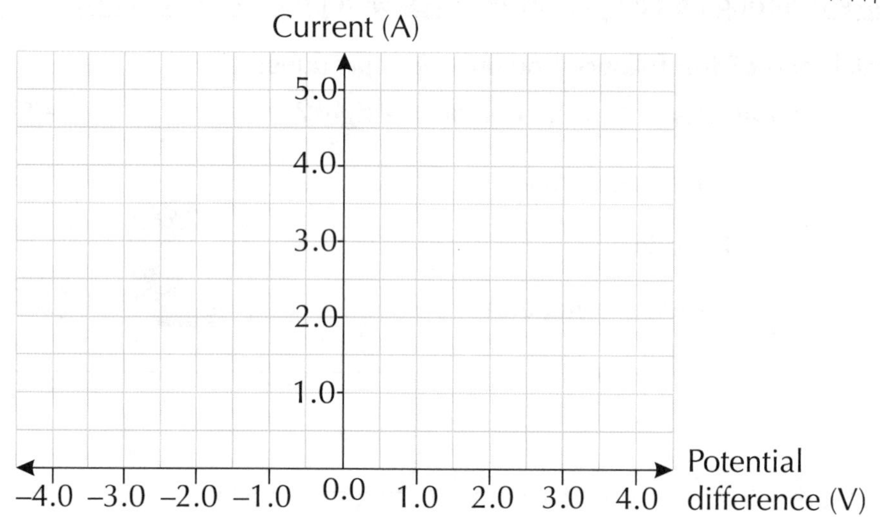

b) Use your *I-V* characteristic to work out what component X is.

Component X is a ....

c) Explain why the current is zero when the potential difference across component X is negative.

....

....

....

I had a straight line I-V graph for a diode — it was very out of character...

Be sure to learn the different *I-V* characteristics. Whether it's for a bulb, a resistor or a diode, they're worth knowing.

Series and Parallel Circuits

It's time to test your knowledge of series and parallel circuits. Stay switched on.

Warm-Up

You can connect components in series or in parallel. How you connect them affects the potential difference, current and resistance throughout the circuit.

There are two diagrams below. One shows a thermistor and an LDR connected in series, the other shows them connected in parallel.
Show which is which by circling the correct word below the diagram.

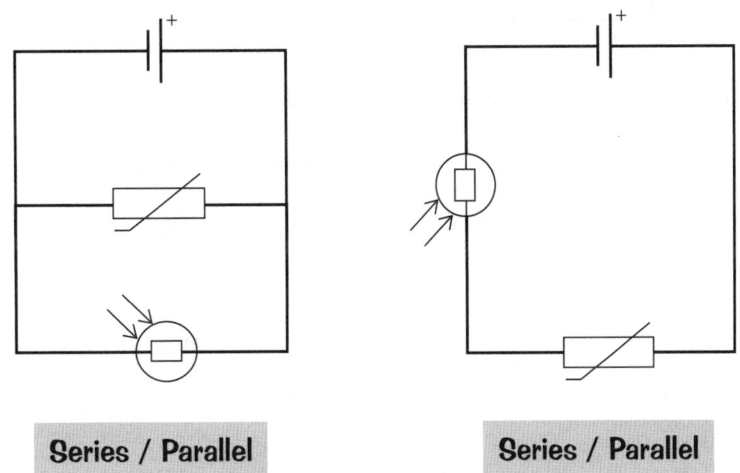

Q1 Two bulbs, 1 and 2, are set up in a circuit with an ammeter and a voltmeter. This is shown in **Figure 1**. The battery supplies a potential difference of 8 V to the circuit.

Figure 1

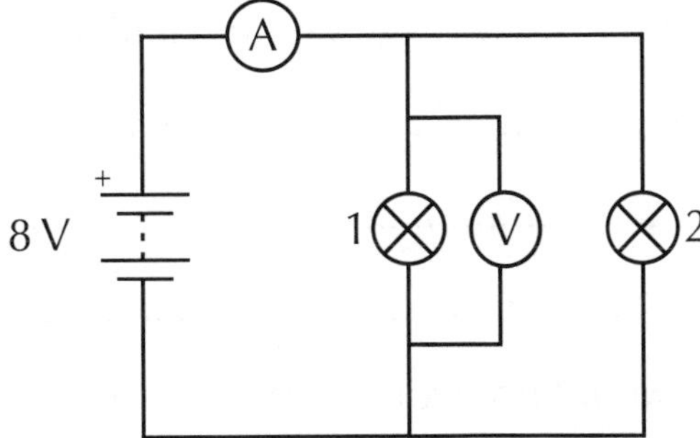

a) What would you expect the reading on the voltmeter to be?

voltmeter reading = ... V

b) The current flowing through the ammeter is 8 A.
The current flowing through bulb 1 is 3 A.
What is the current flowing through bulb 2? Circle the correct answer below.

3 A   5 A   8 A   11 A

Q2 **Figure 2** shows a series circuit. The circuit contains two resistors, R_1 and R_2. R_1 has a resistance of 2 Ω. R_2 has an unknown resistance. Three cells supply a potential difference of 2 V each. An ammeter records a current of 0.5 A in the circuit.

Figure 2

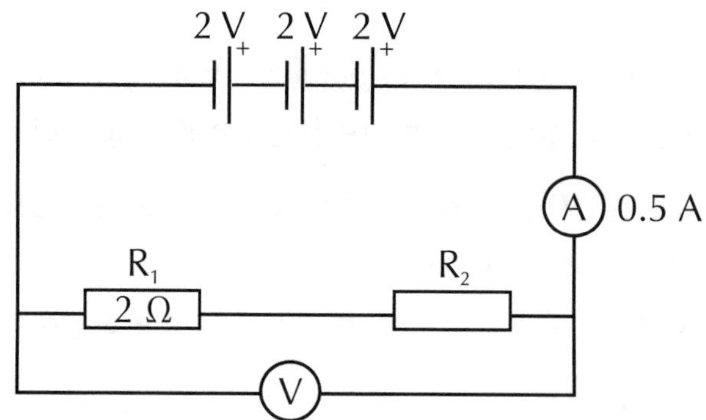

a) What is the total potential difference provided by the cells?

potential difference = ... V

b) Calculate the total resistance of the circuit. Give your answer to 2 significant figures.

Remember, $V = IR$.

total resistance = ... Ω

c) What is the resistance of resistor R_2? Tick **one**.

☐ 6 Ω ☐ 8 Ω ☐ 10 Ω ☐ 12 Ω

Q3 A set of Christmas tree lights is designed to work on the 230 V mains supply. The bulbs in the set of lights are connected in series. Each bulb will only work if there is a potential difference of around 12 V across it.

a) Suggest **one** way you could check to make sure that these bulbs are connected in series and not in parallel.

....

....

....

b) Explain why the bulbs would not work properly if they were connected in parallel and plugged into the mains supply.

....

....

....

There's no denying it — series circuits are unparalleled...

Like apples and oranges, series and parallel circuits are very different from each other. Unless you're hungry, you only need to focus on series and parallel circuits here. Learn the differences between them and life shouldn't be too bad.

Electricity in the Home and Power

Without electricity at home you couldn't charge your phone, toothbrush, robot dog...

Warm-Up

The UK mains supply provides alternating current (ac).

Most electrical appliances are connected to the UK mains supply by three-core cables.

Complete the table below for the wires in a three-core cable.

Wire	Colour(s)	Potential difference (V)
Neutral	Blue	0
Live	Brown	(about) 230
Earth	Green yellow	0

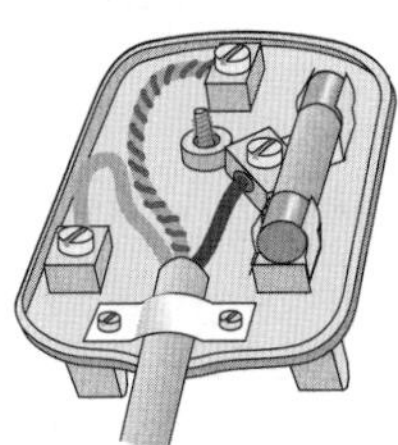

Q1 **Figure 1** shows two appliances, A and B. Appliance A has a plastic casing and appliance B has a metal casing. Plastic doesn't conduct electricity.

Figure 1

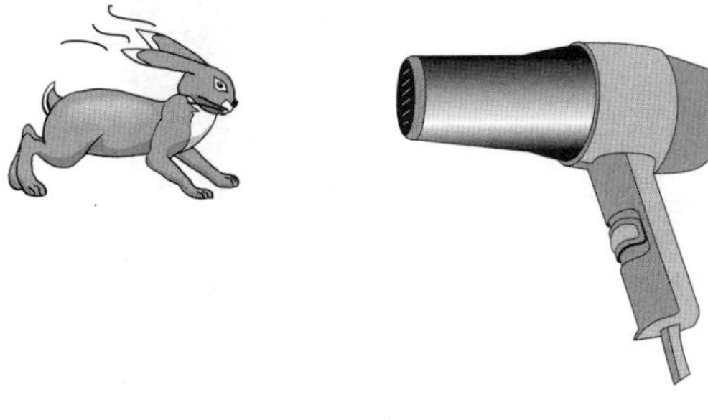

A

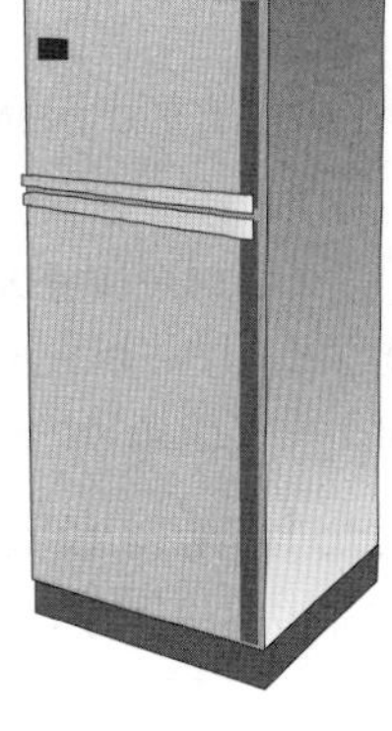

B

a) One appliance has an earth wire connected to its casing.
The other appliance does not have an earth wire.
Circle the letter of the appliance in **Figure 1** that does not require an earth wire.

b) Appliance A is plugged into the UK mains supply and the motor inside appliance A turns. Complete the passage by using some of the words below.

chemical	kinetic	mechanically	electrically

Energy is transferred .... from the mains supply to the .... energy store of the motor in appliance A.

Q2 **Figure 2** shows two graphs of potential difference against time for two different circuits.

Figure 2

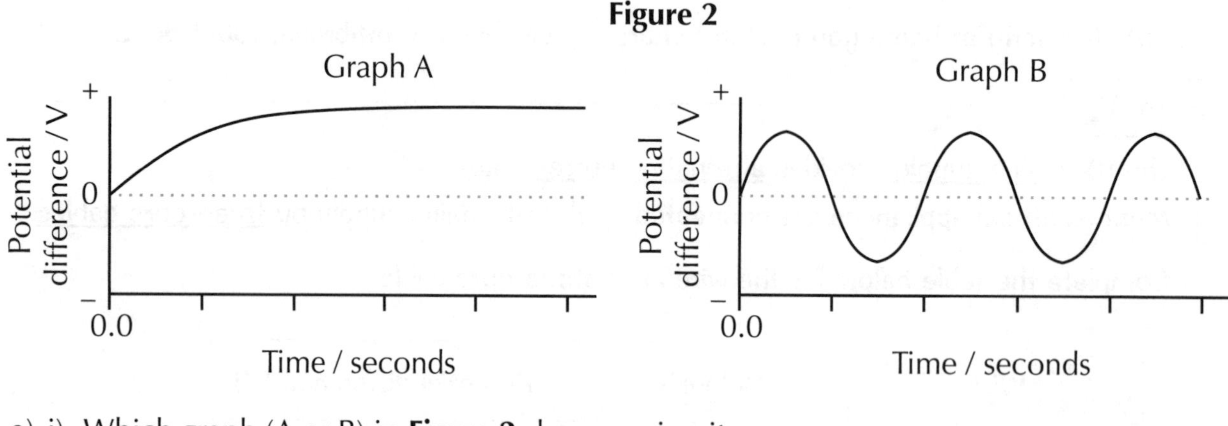

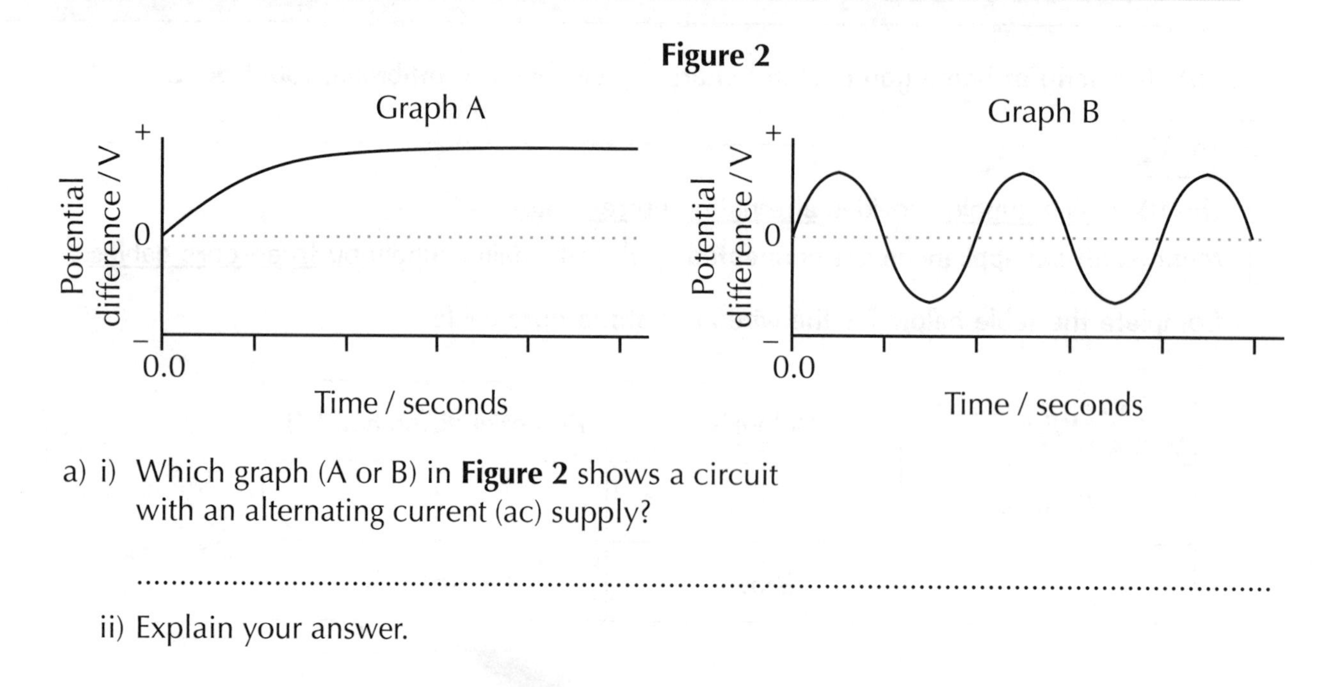

a) i) Which graph (A or B) in **Figure 2** shows a circuit with an alternating current (ac) supply?

....

ii) Explain your answer.

....

....

b) Which graph (A or B) in **Figure 2** could be showing the potential difference in the live wire of a three-core cable?

....

Q3 Gerry is shopping for a new hairdryer.

Gerry has a choice between two hairdryers — the Blaster and the Zoomer. The power of the Blaster is 810 W and the power of the Zoomer is 1500 W.

a) Calculate how long it takes the Zoomer to transfer 9.0×10^5 J of energy.

time = ... s

b) Calculate how much energy the Blaster transfers in the same amount of time.

energy = ... J

Uh oh — Live and Earth went on a date and they really connected...

Alternating current can't make up its mind, always changing direction. Direct current never changes direction.

More on Power

They say knowledge is power — and this knowledge is all about power...

Warm-Up

The <u>power</u> of a device depends on the <u>potential difference</u> across the device and the <u>current</u> through it. What equation links power, potential difference and current?

....

Q1 A circuit in a laptop carries a current of 2.9 A. The resistance of the circuit is 3.8 Ω.

a) Circle the equation that links power, current and resistance.

$P^2 = IR^2$ $\quad$ $P = I^2R$ $\quad$ $P = IR^2$ $\quad$ $P = I^2R^2$

b) Use this equation to calculate the power in the laptop's circuit.

power = ... W

c) Use your answer to b) to calculate the potential difference of the laptop's circuit.

potential difference = ... V

Q2 **Table 1** contains data for four different appliances plugged into the mains supply. It shows the power rating of each appliance and the current flowing through each appliance. The potential difference across each appliance is 230 V. Complete **Table 1**. Give your answers to 2 significant figures.

Table 1

Appliance	Power (W)	Current (A)
Washing machine		3.1
Lamp		0.17
Kettle	2600	
Ceiling fan	78	

Q3 An electric blanket uses 21 V to heat up when it is switched on. The blanket's circuit has a power of 50 W.

a) Calculate the current through the electric blanket's circuit.

current = ... A

b) Calculate the resistance of the circuit in the electric blanket.

resistance = ... Ω

Q4 Oscar has lost his keys so he uses a torch to find them. The torch has a 1.5 V battery. Five minutes after switching the torch on, Oscar finds his keys. In this time, the torch transfers 1.43 kJ of energy.

a) Calculate the total charge that passes through the torch while Oscar is looking for his keys.

charge = ... C

b) The brightness of the torch bulb increases if the power of the torch increases. The 1.5 V battery is replaced with a 3 V battery. Explain how this would affect the brightness of the torch. You can assume the current remains constant.

....

....

....

....

....

You got the power...

Phew, that was a lot of equations. Double-check that you have rearranged the equations correctly and used the correct units in your working. Keep practising this and you'll be powering through any new questions in no time at all.

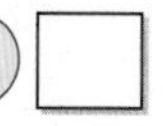

The National Grid

You're powering through — now it's time to step up and test your knowledge of the national grid.

Warm-Up

Electricity is distributed via the national grid.
Transformers are very important for power transmission in the national grid.

Complete the passage using the words given below.

increase	decrease	step-up	step-down

.... transformers are used near power stations to .... the potential difference. The electricity is then transmitted through cables to near homes and factories where .... transformers .... the potential difference.

Q1 A transformer in the national grid changes a potential difference from 11 000 V to 230 V.

a) What type of transformer is this?

....

b) Complete the sentence below by circling the correct word or phrase.

This transformer **increases / decreases / doesn't affect** the current.

Q2 Electrical power is transferred using the national grid.
Tick the **one** statement below that is incorrect.

☐ **A** Current is kept as low as possible when energy is transferred between transformers.

☐ **B** Potential difference is increased at a step-up transformer.

☐ **C** Current is increased before being transmitted through overhead cables.

☐ **D** Potential difference is decreased by a transformer so energy can be used by consumers.

Time to step down from my position as CGP jokes writer I think...

So turning on the TV is as easy as a power station producing electricity which is transformed and carried all around the country in a giant web of high-voltage cables, before being transformed again and used by consumers (like you). Phew.

The Particle Model and Motion in Gases

The particle model describes how particles move, and how they're arranged, in different materials.

Warm-Up

A substance can be a <u>solid</u>, a <u>liquid</u>, or a <u>gas</u>. These are the <u>three states of matter</u>.

Draw <u>lines</u> to match each state of matter to the correct description of its particles.

GAS	The particles are close together but can move past each other easily.
LIQUID	There are no forces between the particles so they aren't held together.
SOLID	Particles are held close together by strong forces.

Q1 Miroslav sells cylinders of helium gas for balloons in his shop. The cylinders have a constant, fixed volume. At night, the temperature of the shop and the cylinders decreases.

Use words from the box to complete the description below.

kinetic	faster	decreases	more
slower	increases	chemical	less

When the temperature of the cylinders decreases, the average energy in the .... energy stores of the helium particles decreases. This means the particles move .... and collide with the walls of the container .... often. So the pressure of the helium gas ....

Keep your cool — it'll relieve some of the pressure...

The particle model helps to explain how temperature, pressure and energy in kinetic energy stores are all related. Make sure you understand how it all works so you can cope with whatever's thrown at you, without getting hot under the collar.

Density of Materials

Density is all about how compact stuff is. It sounds pretty dull, but it's actually super useful.

Warm-Up

Solids, liquids and gases all have different densities.
Density is a measure of the amount of mass in a given volume.

Which of the following is the correct equation for density?

- ☐ density = mass × volume
- ☐ density = mass ÷ volume
- ☐ density = volume ÷ mass
- ☐ density = mass × height

PRACTICAL

Q1 Chloe wants to do an experiment to find the density of a toy.

a) Which of the following would be the most suitable piece of equipment to measure the mass of the toy?

- ☐ **A** A newtonmeter that measures to the nearest newton.
- ☐ **B** A balance that measures to the nearest kilogram.
- ☐ **C** A balance that measures to the nearest gram.
- ☐ **D** A measuring cylinder that measures to the nearest (centimetre)^3.

Chloe has a beaker containing some water. She places the toy into the beaker. **Figure 1** shows the beaker before and after the toy was placed inside.

Figure 1

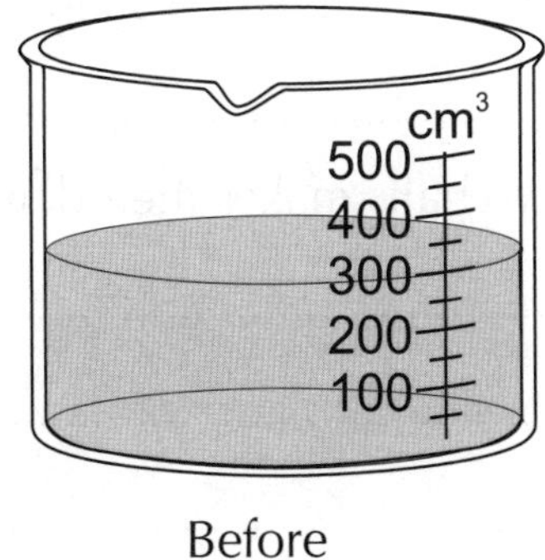

Before

After

When the toy was placed into the beaker, the level of the water increased.
The toy pushed a volume of water out of the way that was equal to the toy's volume.

b) What is the volume of the toy?

volume = ... cm^3

Q2 Katie is building a garden shed using planks of wood.
One plank of wood has a mass of 6.7 kg and a volume of 0.015 m^3.

a) Calculate the density of the plank.

density = ... kg/m^3

b) Katie accidently steps on a plank of wood and snaps it into a big piece and a small piece. Which **one** of the following statements is correct?

☐ **A** The density of the big piece is greater than the density of the small piece.
☐ **B** The density of the big piece is the same as the density of the small piece.
☐ **C** The density of the big piece is lower than the density of the small piece.

Q3 Suneha is a judge for a juggling contest. She is checking all the juggling balls are fair.

a) Suneha measures a juggling ball to have a volume of 6.5×10^{-5} m^3.
The juggling ball has a density of 1800 kg/m^3.
Calculate the mass of the ball.

mass = ... kg

b) The juggling balls must be solid balls. One of the balls Suneha checks is a hollow ball filled with air. The particle arrangement inside the hollow ball is different to the particle arrangement in a solid ball.

i) How will the density of the hollow ball be different to that of a solid ball?

....

ii) Explain how the particle arrangement in the two balls makes their densities different.

....

....

....

....

....

My brother says I'm pretty dense ...

Density is a key concept to understand in GCSE Science. Make sure you know how to calculate density, mass and volume, and how the particle model can explain differences in density and the three states of matter.

Internal Energy, State Changes and Latent Heat

Specific latent heat is the first thing I worry about when my chocolate's melted all over my bag...

Warm-Up

A <u>change of state</u> (e.g. a solid becoming a liquid) can happen when a substance is <u>cooled</u> or <u>heated</u>.

Which of the following is the energy required to change 1 kg of a <u>liquid</u> at its boiling point into a <u>gas</u>? Tick one.

- ☐ the specific heat capacity
- ☐ the specific latent heat of vaporisation
- ☐ the specific latent heat of fusion

Q1 A cauldron of pure water is heated over a fire. When the water reaches 100 °C, it stays at 100 °C, even though it is still being heated. Which sentence below correctly describes what is happening to the water? Tick **one** box.

- ☐ **A** The internal energy of the water is not changing.
- ☐ **B** The water is cooling down.
- ☐ **C** The energy supplied by the fire is being used to change the state of the water into water vapour.

Q2 A kettle is switched on so that it supplies energy to the water inside it.

Use the correct words and phrases from the box to fill in the gaps in the sentences below, to explain what is happening to the water.

boil	break	stay constant	internal
increase	form	condense	magnetic

The energy supplied by the kettle increases the water's .... energy.

This causes the temperature of the water to ....

After the temperature gets high enough, the water begins to ....

The energy being transferred is now being used to .... bonds between particles. The temperature will .... during this process.

PRACTICAL

Q3 Pandita heats a solid block of wax. The starting temperature of the block was 20 °C. She measures the temperature of the wax each minute. **Figure 1** shows her results.

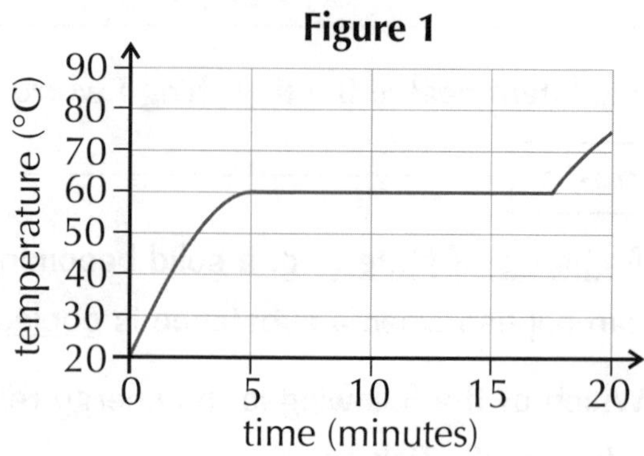

a) i) At what temperature does the wax become liquid?

Temperature = ... °C

ii) Circle the word that describes what is happening to the wax at this temperature.

Freezing Melting Sublimating Boiling

b) Pandita measures the mass of wax at the end of her experiment. She finds the mass to be 1 kg. What was the mass of wax at the start of the experiment? Explain your answer.

....

....

Q4 Dave is doing some experiments with ice. Ice has a melting point of 0 °C. It has a specific latent heat of fusion of 334 000 J/kg.

a) Dave uses a heater to completely melt some ice at 0 °C. The heater transfers 520 000 J of energy to melt the ice. Calculate the mass of the ice.

mass of ice = ... kg

b) Dave puts 32 g of ice at 0 °C into a beaker of water. The ice melts. Calculate the energy transferred to the ice by the water as the ice melts.

Don't forget to convert the units from g to kg.

energy transferred = ... J

So long boiled water — you will be mist...

Be careful not to confuse specific latent heat (the energy needed to change the state of a 1 kg mass) with specific heat capacity (the amount of energy needed to raise the temperature of 1 kg of a substance by 1 °C).

Developing the Model of the Atom

Be wary. You just can't trust atoms — they make up everything...

Warm-Up

The diagram shows the current model of the atom.
This is what scientists believe the atom looks like.

Complete the diagram so that it shows the name and location of the three particles that make up the atom.

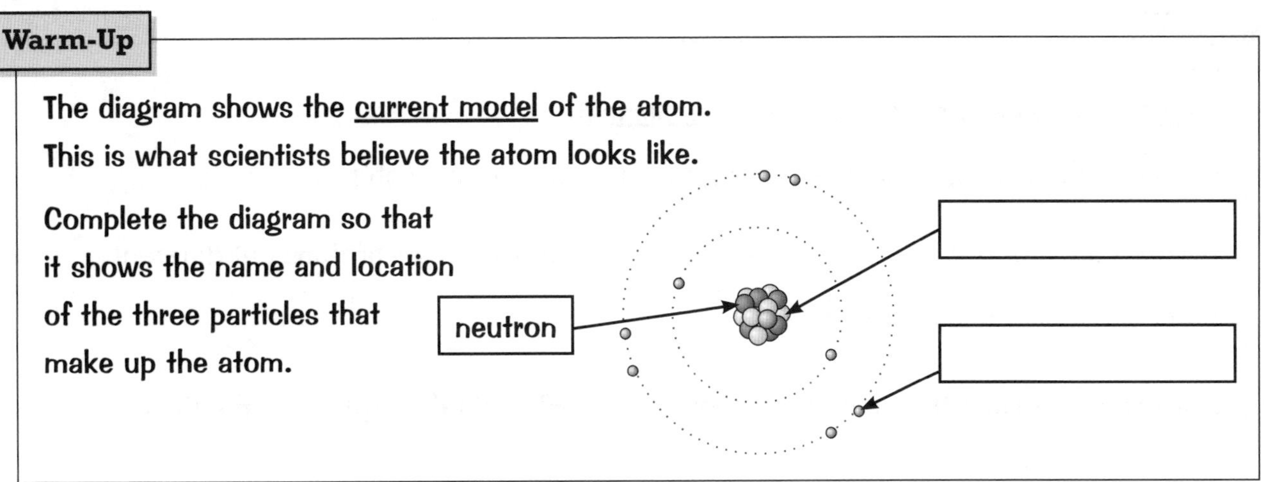

Q1 **Figure 1** shows the path of a positively charged alpha particle hitting a piece of thin gold foil. Opposite charges attract, and like charges repel.

Figure 1

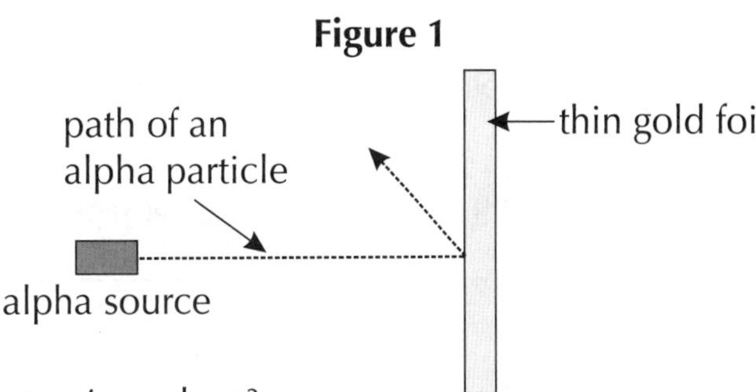

a) What does the path of this alpha particle suggest about the charge of an atom's nucleus? Tick the box next to the correct answer.

☐ it has a positive charge ☐ it has a negative charge ☐ it is uncharged

b) Explain your answer.

....

....

We've struck gold!

Q2 A hydrogen atom contains just one electron. The electron orbits the nucleus in the lowest energy level. The electron absorbs enough radiation to allow it to move to the next energy level within the atom. State how the radius of the electron's orbit has changed. Explain your answer.

....

....

My head's like an atom — mostly made up of empty space...

At first we thought an atom was just one particle, then we realised it was made up of different particles and came up with the plum pudding model, followed by the nuclear model, then the Bohr model... who knows what it will look like in the future.

Isotopes, Radiation and Nuclear Equations

Alpha, beta and gamma — they're totally rad(-iation)...

Warm-Up

Isotopes are different forms of the same element.

Complete the passage below by circling the correct word in each pair.

Isotopes are atoms which have the same number of **neutrons / protons** but different numbers of **neutrons / protons** .

Draw lines to match the type of radiation on the left to its description on the right.

Alpha particle	An electron.
Beta particle	A type of electromagnetic radiation.
Gamma ray	2 neutrons and 2 protons — the same as a helium nucleus.

Q1 For each of the following isotopes, state the number of protons and the number of neutrons.

a) $^{3}_{1}H$

Protons: ....

Neutrons: ....

b) $^{14}_{6}C$

Protons: ....

Neutrons: ....

c) $^{14}_{7}N$

Protons: ....

Neutrons: ....

Q2 Three sources, X, Y and Z, emit either alpha, beta or gamma radiation. Each source only emits one type of radiation. **Table 1** shows the range in air for the radiation emitted by each source.

Complete **Table 1** to show the type of radiation emitted by each source.

Table 1

Source	Range in air / m	Type of radiation emitted
X	50	
Y	0.03	
Z	1	

Q3 Nuclear equations are one way of showing radioactive decays.

a) Complete the nuclear equation below to show an atom of radon-222 ($^{222}_{86}Rn$) emitting a gamma ray.

$$^{222}_{86}\text{Rn} \longrightarrow \;^{\cdots\cdots}_{\cdots\cdots}\text{Rn} + \;^{\cdots\cdots}_{\cdots\cdots}\gamma$$

b) An atom of thorium-234 ($^{234}_{90}Th$) emits a beta particle and becomes an atom of protactinium (Pa). Write a nuclear equation to show this decay.

....

Q4 **Figure 1** shows beta radiation being used to control the thickness of a sheet of metal being made in a factory.

Figure 1

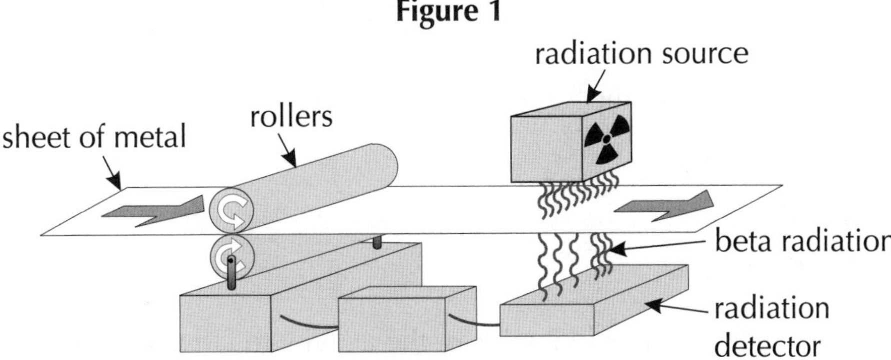

A radiation detector and the radiation source are placed on opposite sides of the metal sheet. If too little radiation reaches the radiation detector through the metal, the rollers are adjusted to make the sheet thinner. If too much radiation is detected, the rollers are adjusted to make the sheet thicker.

a) Explain why alpha radiation wouldn't be suitable for this use.

....

....

b) The beta radiation source is replaced with a gamma radiation source. Will the machine still be able to control the thickness of the metal sheet? Explain your answer.

....

....

....

Radiation — as easy as alpha, beta, gamma...

You could get asked about different uses of radiation, like in question 4. Keep your cool and look at what properties you want the radiation being used to have, then compare that with what you know about alpha, beta and gamma radiation.

Half-life

Half-life is not the zombie-like feeling you get when the alarm rings in the morning. It's to do with measuring the rate at which unstable nuclei decay. Thrilling stuff.

Warm-Up

Half-life is the time taken for the number of radioactive nuclei in an isotope sample to halve.

Only one of the following statements is true. Tick the box next to the correct one.

☐ **A** The half-life of a radioactive isotope will always be the same.

☐ **B** You can predict the order in which nuclei in a sample will decay.

☐ **C** Radioactive substances give out radiation once they have been activated.

Q1 The initial activity of a sample of a radioactive isotope is 8000 Bq. After 80 seconds, the activity of the sample is 2000 Bq.

a) What is the half-life of the sample? Tick **one** box.

☐ **A** 20 seconds

☐ **B** 40 seconds

☐ **C** 80 seconds

b) How many half-lives would it take for the activity to fall from 8000 Bq to below 600 Bq? Give your answer as a whole number.

number of half-lives = ...

Q2 **Table 1** shows the count rate from a radioactive sample.

Table 1

Time / mins	0	5.0	10.0	15.0	20.0
Count rate / cpm	6200	4380	3100	2190	1550

The sample is said to be 'safe' when the count rate is below 20 cpm. Use calculations to show that the count rate will **not** have reached this 'safe' level after 60 minutes.

....

....

....

Q3 Sample A is a pure sample of a radioactive isotope. The activity of sample A decreases with time. This is shown in **Figure 1**.

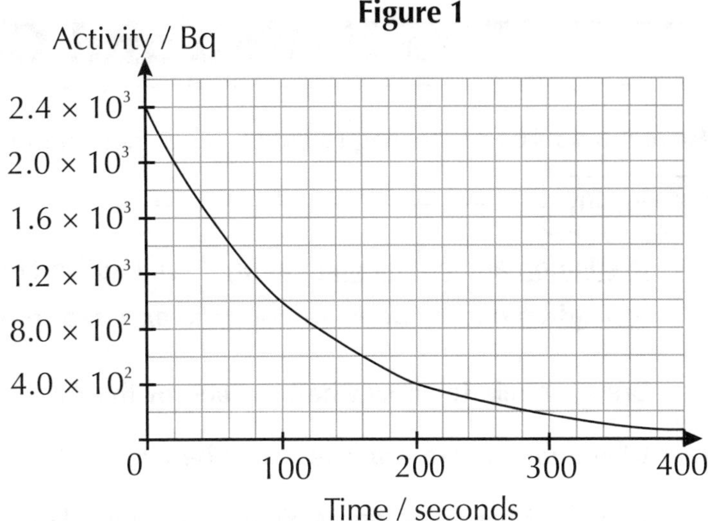

a) What is the half-life of sample A?

half-life = ... s

b) i) The initial activity of sample A was 2.4×10^3 Bq.
What was the activity of the sample after 3 half-lives?

activity = ... Bq

ii) How long did it take for the activity to fall to the value you calculated in b) i)?

time = ... s

c) The initial mass of sample A was 28 g.
What mass of the radioactive isotope was left after 2 half-lives?

mass = ... g

d) Sample B has the same initial activity as sample A. However, the radioactive isotope in sample B has a much shorter half-life than the isotope in sample A.

Look at **Figure 2**. One of the lines, X, Y or Z, correctly shows how the activity of sample B varies with time. Which one?

Line = ...

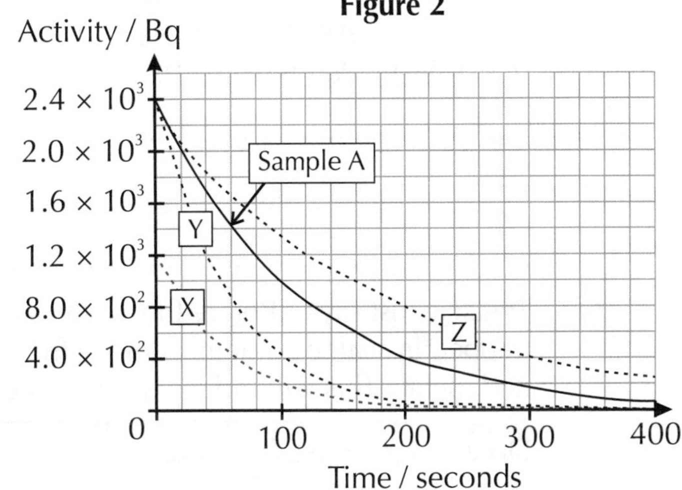

Maya's activity = 0

Half-life — half the fun of real life...

Remember, half-life is the time it takes for the activity of a radioactive sample to halve. The half-life of a radioactive sample is always the same — it doesn't matter what activity you start with or how big the sample is.

Irradiation and Contamination

Having a source of ionising radiation on or near you can be very dangerous...

Warm-Up

Irradiation and contamination by a radioactive source can damage the human body. Irradiation occurs when an object is exposed to nuclear radiation.

Draw a circle to show whether each of the following sentences are true or false.

When objects are irradiated they become radioactive.	True	False
An object has to be contaminated to be irradiated.	True	False
Objects don't have to touch a radioactive source to be irradiated by it.	True	False

What is contamination?

....

....

....

Q1 Radioactive waste emits ionising radiation and has to be carefully disposed of.

a) A scientist suggests burying the waste hundreds of metres underground.

Explain **one** way that burying the radioactive waste would reduce the risk of people being irradiated by it.

....

....

....

b) The scientist says, "Before the waste is buried, it should be trapped in sealed containers made from thick layers of glass and metal. This will reduce the risk of contamination.".

Do you agree? Explain your answer.

....

....

....

Q2 **Figure 1** shows two scientists handling samples of radioactive material just before doing an experiment using the sources.

Figure 1

a) How should the radioactive samples be stored when not in use? Explain your answer.

b) One of the scientists is taking sensible safety precautions, but the other is not. Using **Figure 1**, describe **two** things which the careless scientist is doing wrong and how they increase the risk of irradiation or contamination (or both).

What the scientist is doing wrong:

How this increases the risk of irradiation or contamination:

What the scientist is doing wrong:

How this increases the risk of irradiation or contamination:

c) The careless scientist's hands become contaminated by some of the radioactive source they have been using. They then eat a sandwich for lunch without decontaminating their hands. Explain why this is dangerous.

Radiating positive physics vibes...

Irradiation and contamination have to be taken very seriously to avoid causing people unnecessary harm. Make sure you know which one is which and why they're dangerous — it could come in handy for GCSE Science.

Contact and Non-Contact Forces

If you're struggling with forces, this page will give you a push (or pull) in the right direction...

Warm-Up

Quantities in physics can be vectors or scalars. Forces are vector quantities.

Draw a line to match each type of quantity to its description.

Vector quantities	Have magnitude but no direction.
Scalar quantities	Have magnitude and a direction.

If two objects have to be touching for a force to act, the force is a contact force.
If the objects don't need to be touching for a force to act, it's a non-contact force.

Circle the one contact force shown below.

gravitational force   friction   electrostatic force   magnetic force

Q1 The force diagram in **Figure 1** shows a flamingo standing on one leg. Two forces, A and B, act on the flamingo. The two forces are represented by the arrows A and B in **Figure 1**.

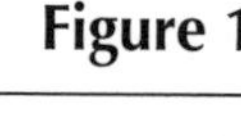

Figure 1

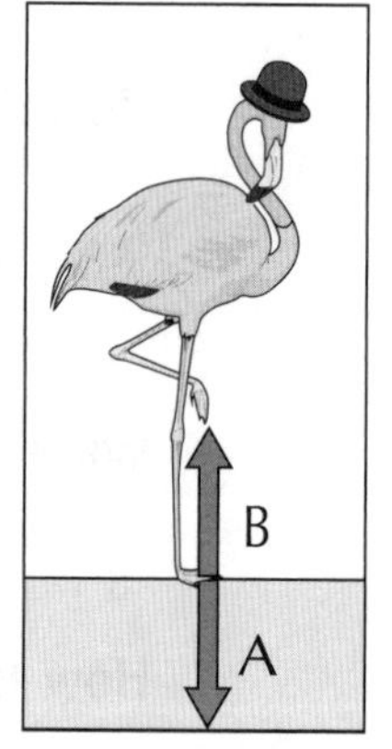

a) i) What type of force is force A? Tick **one** box.

- [ ] **A** gravitational force
- [ ] **B** friction
- [ ] **C** normal contact force
- [ ] **D** tension

ii) Is force A a contact or a non-contact force?

....

b) What does the length of the arrows in **Figure 1** represent?

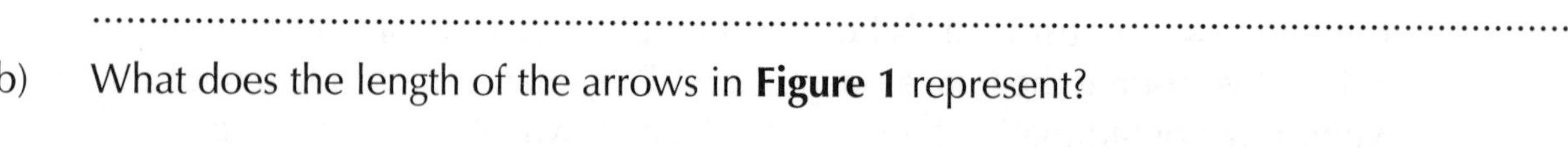

....

c) Are forces A and B the same size or different sizes? Circle the correct answer.

same size   different sizes

I had a heated chat about contact forces — there was tension in the room...

Contact forces can't act unless two objects are touching each other. Air resistance is a contact force. In this case, an object has to touch the air for air resistance to act on it — if there's no air, then there's no force.

Weight, Mass and Gravity

Physics, gravity — they both just bring me down...

Warm-Up

All objects have a mass.

Which of the following statements about mass is not true?
Tick one box.

- ☐ Mass is measured in kg.
- ☐ An object's mass changes depending on where in the universe it is.
- ☐ Mass is the amount of matter (stuff) in an object.

The weight of an object at a certain point depends on its mass and the gravitational field strength at that point.

What equation links weight, mass and gravitational field strength?

....

Q1 A student is measuring the weights of objects in a classroom experiment.
She hangs a 2.0 kg object from a newtonmeter.
The gravitational field strength in the classroom is 9.8 N/kg.

a) Calculate the weight of the object.

weight = ... N

b) Which of the following statements is true? Tick **one** box.

- ☐ **A** The object's weight is directly proportional to the object's mass.
- ☐ **B** The object's weight is inversely proportional to the object's mass.
- ☐ **C** The object's weight is equal to the object's mass.
- ☐ **D** The object's weight is equal to the object's centre of mass.

c) The Moon's gravitational field strength is weaker than the Earth's.
Underline the correct word or phrase to complete the following sentence.

If the student did the same experiment on the Moon, the object would weigh **less** / **more** / **the same**.

Q2 Ollie has been feeding his dog Peri-Peri a bit too much chicken. He decides to weigh her. Gravitational field strength $g = 9.8$ N/kg on Earth.

a) Ollie finds that Peri-Peri has a weight of 540 N. Calculate Peri-Peri's mass.

mass = ... kg

b) Peri-Peri goes on a diet for three months. After three months, Ollie weighs her again. She now has a mass of 47 kg. By how much has her weight changed?

change in weight = ... N

Q3 A robot lands on the surface of Titan, a moon of Jupiter. It weighs a set of known masses. The results are shown in **Table 1**.

Table 1

Mass (kg)	0.10	0.20	0.30	0.40	0.50
Weight (N)	0.15	0.30	0.42	0.55	0.68

a) Plot the data in **Table 1** on the grid in **Figure 1**. The first two points have been plotted for you.

b) Draw a line of best fit on **Figure 1**.

c) Use **Figure 1** to estimate the gravitational field strength at Titan's surface. Give your answer to 2 significant figures.

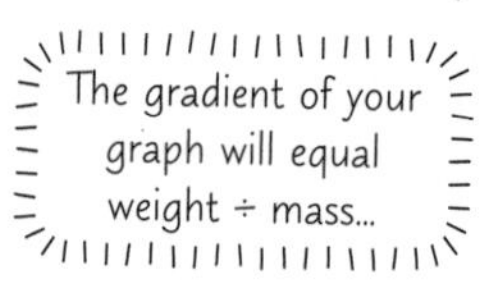

Figure 1

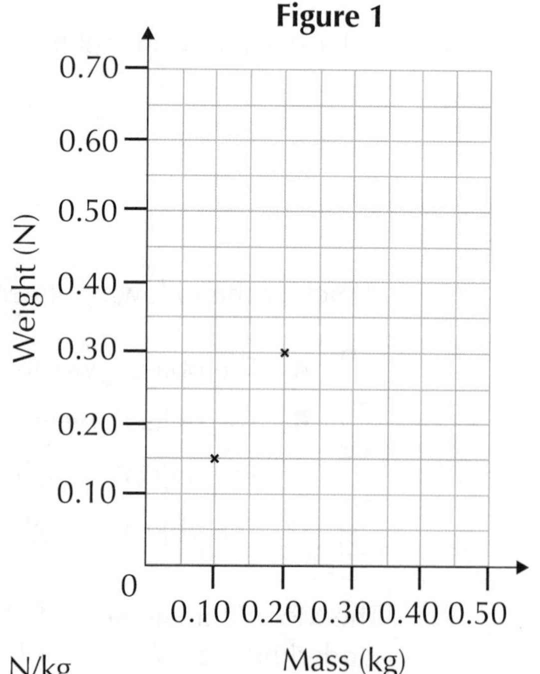

gravitational field strength = ... N/kg

New, totally automated, zero-gravity restaurant — no weight-ers required...

I've discovered how to weigh more but look no different — chill on a big planet like Jupiter. With a gravitational field strength larger than Earth's, weighing yourself might give you a scare but in the mirror you'll look just the same.

Resultant Forces and Work Done

I'm afraid there's no hiding from it — it's time to get some work done on work done.

Warm-Up

Objects usually have more than one force acting on them.

Circle the correct underlined words to complete the sentence below:

If a number of forces act on an object, they can be replaced by a

contact / single / gravitational force called the random / resultant force.

Q1 Simon likes to skateboard. He puts his dog on a lead and gets his dog to pull him along.

a) The dog pulls Simon along with a constant forward force of 365 N.
There is a constant frictional force of 53 N acting on the skateboard.
Calculate the resultant force acting on the skateboard.

resultant force = ... N

b) Use some of the words in the box to fill in the sentences below.

thermal	kinetic	increase	decrease

The work done by the dog against friction causes energy to be transferred

to the .... energy store of the skateboard.

This causes the overall temperature of the skateboard to ....

c) Simon lets his dog off the lead and goes to skateboard on his own.
He pushes off a wall from rest and rolls in a straight line.
There is a constant frictional force of 38 N acting on the skateboard wheels.
Simon slows down and comes to a stop.
The total work done by friction on the skateboard is 525 J.

Calculate the distance that Simon travels between pushing off the wall and coming to a stop.

distance = ... m

24 hour energy transfer — all in a day's work...

Work done is just another way of saying energy transferred. There was a smarty pants at school who, when he had his homework done, used to say he'd got his home-energy transferred. I thought he meant he'd had solar panels installed...

Forces and Elasticity

An object that bends, compresses and stretches — nothing springs to mind...

Warm-Up

If you apply more than one <u>force</u> to an <u>elastic</u> object, like a spring, it will stretch, compress or bend.

Circle the graph below that correctly shows force against extension for an elastic object being stretched.

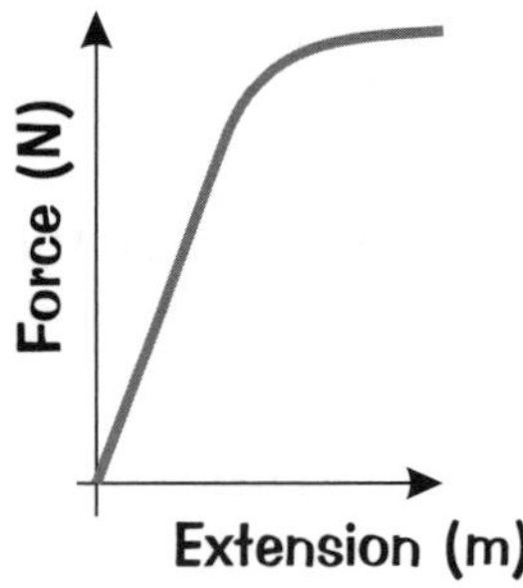

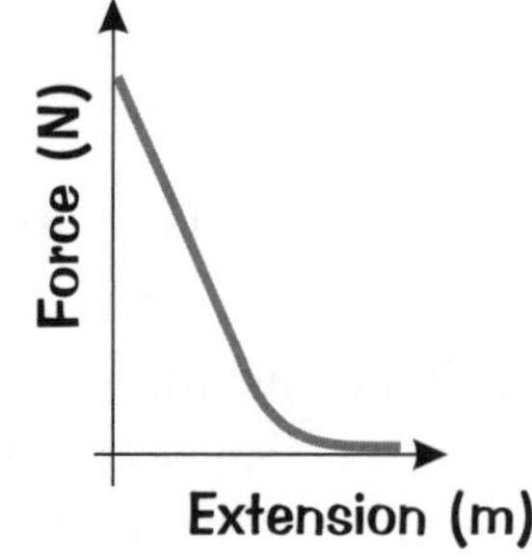

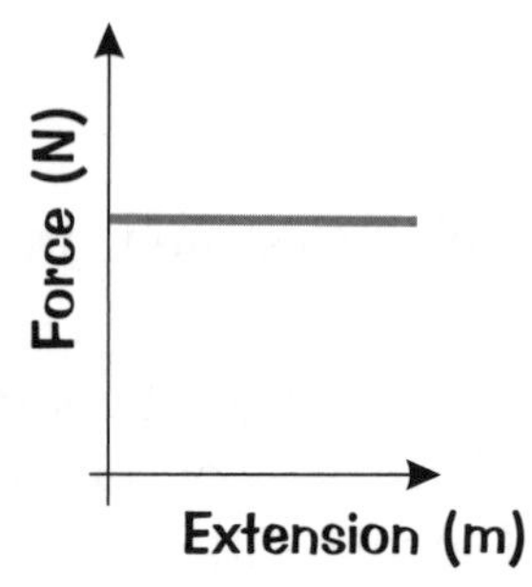

Q1 A fisherman uses a scale to measure the weight of a fish. The scale is shown in **Figure 1**. The spring in the scale is fixed at one end and has a hook at the other.

Figure 1

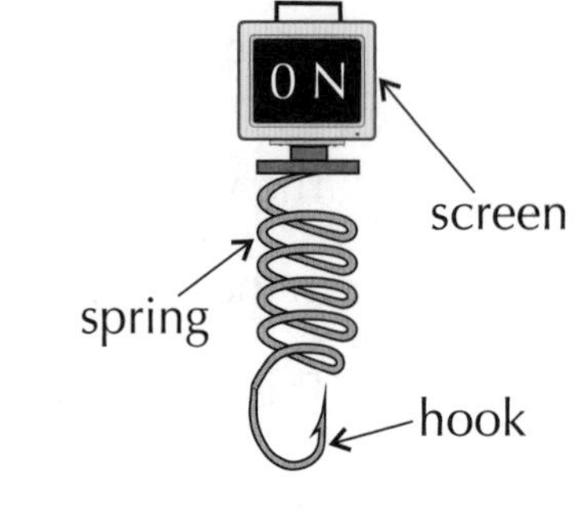

a) What will happen if the spring in the scale is stretched past its limit of proportionality? Tick **one** box.

☐ **A** The spring will snap in half.

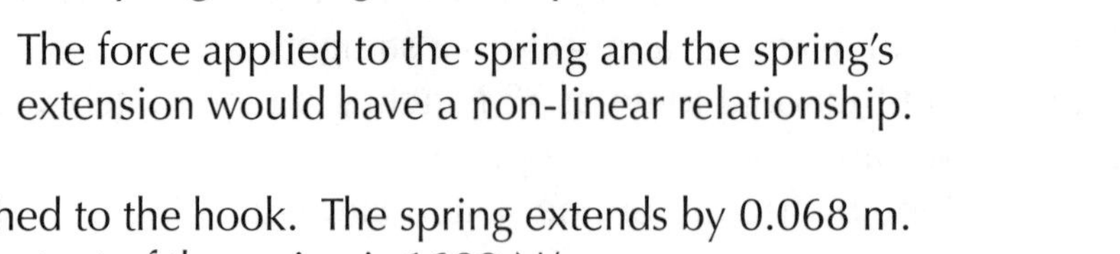

☐ **B** The spring will begin to compress.

☐ **C** The force applied to the spring and the spring's extension would have a non-linear relationship.

A fish is attached to the hook. The spring extends by 0.068 m.
The spring constant of the spring is 1600 N/m.
You can assume the spring does not extend past its limit of proportionality.

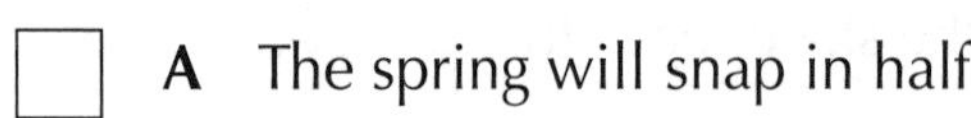

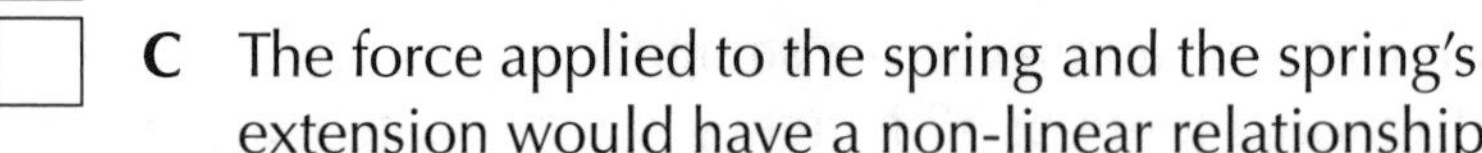

b) i) Circle the equation needed to find the weight, W, of the fish.

$W = k \div e$ $\quad$ $W = ke$ $\quad$ $W = \frac{1}{2}ke^2$ $\quad$ $W = e \div k$

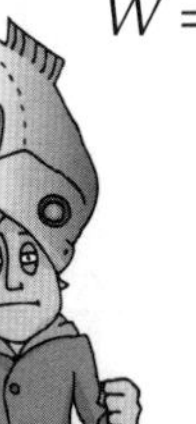

ii) Calculate the weight of the fish.

weight = ... N

PRACTICAL

Q2 Maryam is a bungee jumper who is checking her bungee rope. She carries out an experiment to find how much force it takes to stretch the bungee rope by different amounts. Her results are shown in **Table 1**.

Assume that the bungee rope is always below its limit of proportionality in this question.

Table 1

Extension (m)	Force (N)
3	270
9	750
15	1300
21	1900
27	2500
33	3000

Figure 2

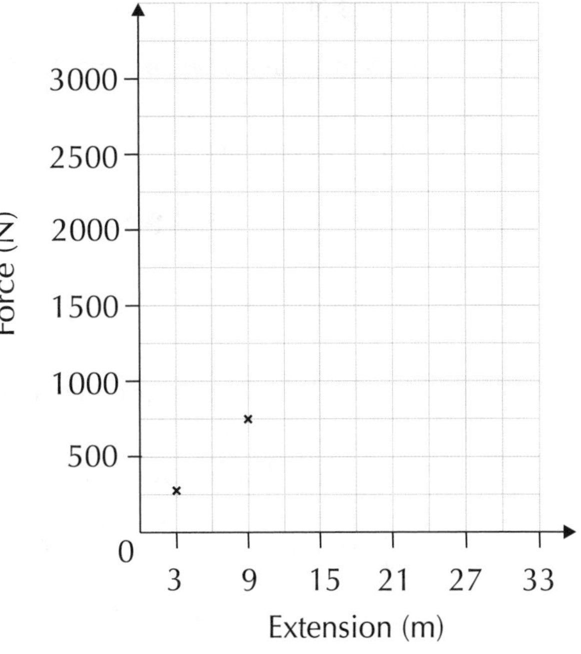

a) Plot the data in **Table 1** and draw a line of best fit on the grid in **Figure 2**. The first two points have been plotted for you.

b) Use your graph to calculate the spring constant of the bungee rope.

The gradient of your graph will equal force ÷ extension...

spring constant = ... N/m

Maryam does a bungee jump using the bungee rope. At the maximum extension of the bungee rope, the energy in the rope's elastic potential energy store is 65 700 J.

c) Using your spring constant from b), calculate the maximum extension of the bungee rope.

extension = ... m

I'm getting an extension in the spring — a conservatory I think...

A spring is just one type of elastic object, so don't get put off if the question involves something different, such as a bungee rope. As long as the object is elastic and below its limit of proportionality, the equations used are just the same.

Distance, Displacement, Speed and Velocity

You may think your physics knowledge is up to speed, but does it have direction..?

Warm-Up

Distance is how far an object moves. Distance is a scalar quantity.
Displacement is the distance an object moves in a particular direction.
As displacement involves direction, it is a vector quantity.

Circle any values below which are displacements:

15 m to the left   900 km/h   15 m

90 miles   18 cm north

Q1 Lois rides her bike to work. It takes her 300 s.

a) i) What would you expect Lois' cycling speed to be? Tick **one** box.

☐ 0.5 m/s   ☐ 6 m/s   ☐ 25 m/s   ☐ 60 m/s

ii) Using your chosen cycling speed, estimate the distance Lois travels to get to work.

distance = ... m

b) One part of Lois' journey is down a hill. It takes her 18 s to cycle 230 m down the hill. What is Lois' average speed down the hill?

speed = ... m/s

Q2 Ealing Broadway train station is about 12 km west of Marble Arch train station.
On one journey, a train takes approximately 1200 s to get from Marble Arch to Ealing Broadway.

Which of the following correctly describes this journey? Tick **one** box.

☐ **A** The average velocity of the train is approximately 10 m/s.
☐ **B** The average velocity of the train is approximately 120 m/s due west.
☐ **C** The average velocity of the train is 0 m/s.
☐ **D** The average velocity of the train is approximately 10 m/s due west.

Q3 A car is on a straight race track. The car travels at a constant speed along the track, from the start line. The track is painted with white lines. The distance between each line is 2 m.

The diagram in **Figure 1** shows two photos of the car on the same part of the track. The car is travelling at 20 m/s.

Figure 1

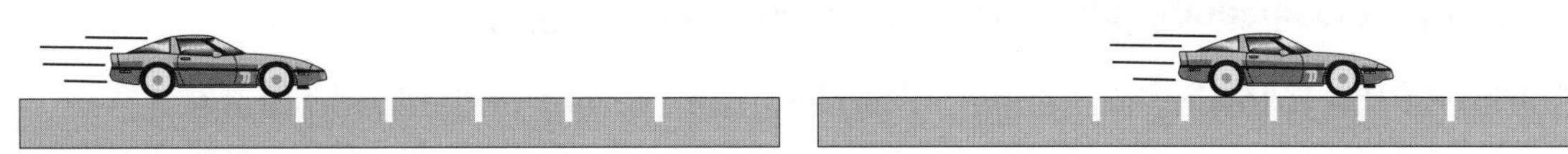

a) Using **Figure 1**, calculate the time between the two photos.

time = ... s

b) When the car reaches the end of the track, it stops and turns around. The car then travels back along the track towards the start line. The car drives along the track at a constant speed of 20 m/s. Which of the following statements is false? Tick **one** box.

- ☐ **A** The car is travelling at the same speed as it was in a).
- ☐ **B** The car will take 10 s to travel 200 m.
- ☐ **C** The car is travelling at the same velocity as it was in a).

c) The car drives on a different race track. Part of the race track curves, as shown in **Figure 2**. The car drives along the track from point A to point B.

Circle the correct phrase to complete the sentence below.

Figure 2

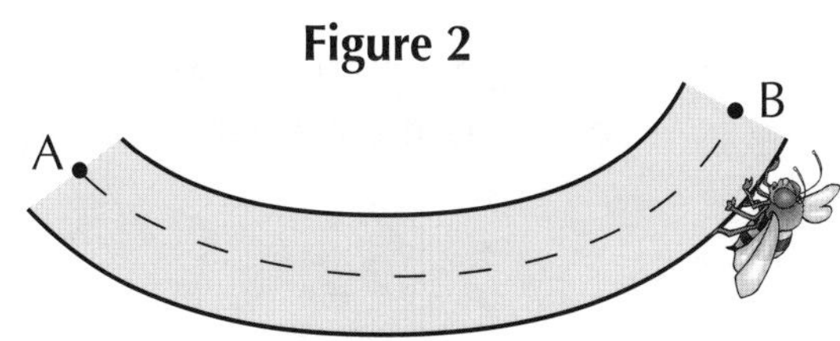

The distance travelled by the car as it drives between points A and B is **bigger than / smaller than / the same as** the displacement of the car.

Q4 It takes the Moon 27 days to complete one orbit. 1 day = 86 400 s. The Moon travels a distance of 2.4×10^9 m in one orbit.

Calculate the average speed of the Moon.

speed = ... m/s

My average speed depends on distance, time, how much I've eaten for tea...

The speed, distance, time formula can be used to calculate an object's constant speed. Yet, in real life, most objects don't travel at a constant speed. For such objects, the formula can be used to calculate an *average* speed instead.

Acceleration

Speeding up whilst jogging is an example of acceleration. It's also very tiring...

Warm-Up

If an object is speeding up or slowing down, it is <u>accelerating</u>.

Complete the following sentence by circling the correct underlined phrase.

Deceleration is when an object <u>speeds up</u> / <u>slows down</u> / <u>doesn't move</u> .

Q1 A runner accelerates from rest to an average running speed in 4 s. He runs in a straight line.

a) What equation links acceleration, a, change in velocity, Δv, and time taken, t? Tick **one** box.

☐ $a = \Delta v \times t$

☐ $a = v \div \Delta t$

☐ $a = v + \Delta t$

☐ $a = \Delta v \div t$

b) Using a typical value for running speed, estimate the runner's acceleration over the 4 s.

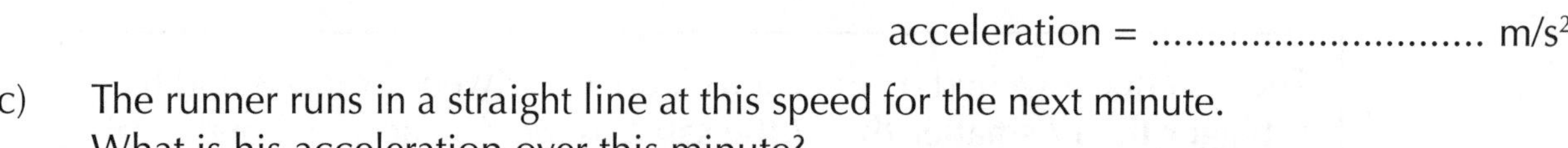

acceleration = ... m/s^2

c) The runner runs in a straight line at this speed for the next minute. What is his acceleration over this minute?

acceleration = ... m/s^2

Q2 A car company produces high speed cars. Their latest car has a maximum acceleration of 3.7 m/s^2.

Calculate the time it takes for the car to accelerate from 12 m/s to 25 m/s when it's moving at its maximum acceleration.

time = ... s

Q3 An egg was dropped from the very top of the Leaning Tower of Pisa. The egg fell freely under gravity. 3.3 s after it was dropped, a chef caught it in her pan.

Any object that is said to be falling freely under gravity only has a gravitational force acting on it.

a) What was the acceleration of the egg as it fell? Tick **one** box.

☐ **A** 0 m/s² ☐ **B** 3.3 m/s² ☐ **C** 4.9 m/s² ☐ **D** 9.8 m/s²

b) At what speed did the egg hit the pan?

speed = ... m/s

c) Calculate how far the egg fell before it was caught.

Remember, there's a list of some physics equations on page 244. You'll need one of these equations to answer this question.

height = ... m

Q4 A tractor decelerates at 1.92 m/s² over 10.82 m, after which its velocity is 3.22 m/s.

a) Calculate the tractor's velocity before it started decelerating.

velocity = ... m/s

b) Calculate how long it takes for the tractor to decelerate over the 10.82 m.

time = ... s

Calculating accelerations — a hobby for those who live in the fast lane...

A favourite page for you calculation-lovers. A tricky part of acceleration calculations is deciding which equation to use. One equation has time and the other has displacement, so check what you're given in the question before rushing in.

Motion Graphs and Terminal Velocity

Graphs — they crop up from (distance-)time to (velocity-)time...

Warm-Up

Speed can be found from the gradient of a distance-time graph.

What can be found from the gradient of a velocity-time graph?

....

An object falling through a fluid will eventually reach a terminal velocity.

Use some of the words below to fill in the following paragraph about terminal velocity:

slow	resultant	gravity	tension	constant	drag

An object falling through a fluid first accelerates due to the force of .... The fluid opposes this force with a drag force. Eventually, the .... force on the object will be zero and the object moves at a .... speed, called its terminal velocity.

Q1 Three trams travel between two points. **Figure 1** shows the distance-time graphs for each tram and a description of each tram's journey.

Draw lines between the graphs and the descriptions to show which description belongs to which graph.

Figure 1

A

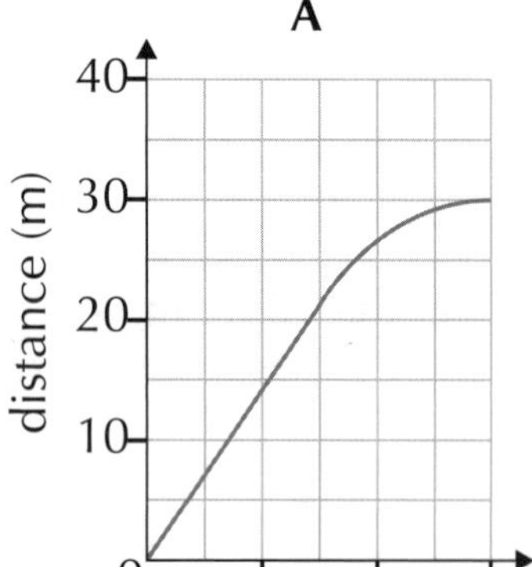

B

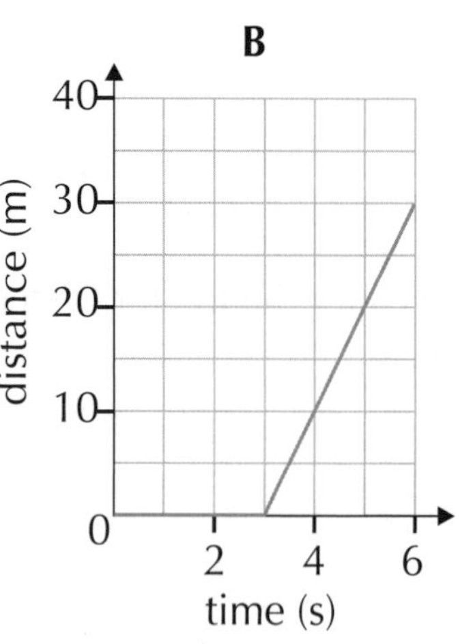

C

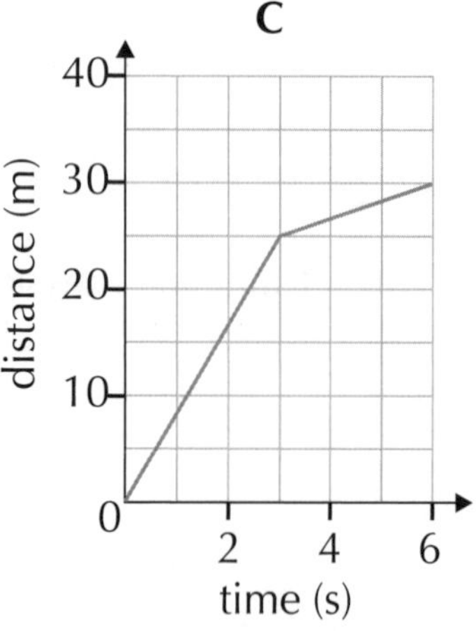

The tram travels at a constant speed for 3 s and then travels at a slower constant speed for the next 3 s.	The tram travels at a constant speed for 3 s and then decelerates for the next 3 s.	The tram remains stationary for 3 s and then travels at a constant speed for the next 3 s.

Q2 A skydiver, with a weight of 590 N, falls at terminal velocity.
The only two forces acting on the skydiver are weight and air resistance.

A student says: "At terminal velocity, the air resistance force acting on the skydiver is 590 N." Is this true or false? Circle the correct answer.

True　　　　　　　　False

Q3 Mandip walks from her classroom to the playing field for football training.
She arrives, waits 20 s, then realises she's left her boots in the classroom.
She walks back to the classroom at the same speed. She takes 60 s to look for her boots and then runs back to the playing field at twice her walking speed.
The distance between the classroom and the playing field is 200 m.

Figure 2 shows an incomplete distance-time graph for Mandip's journey.

Figure 2

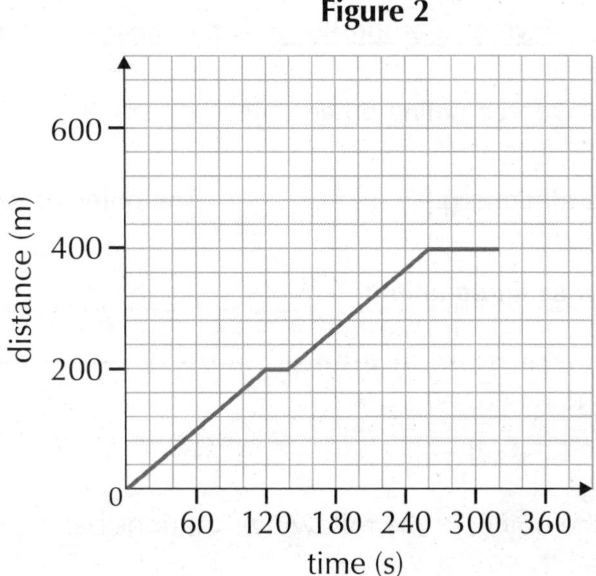

a) How long did it take Mandip to walk to the playing field initially? Circle **one** answer.

120 s　　　　130 s　　　　140 s　　　　200 s

b) Calculate Mandip's speed when she walked to the playing field initially.
Give your answer to 2 significant figures.

speed = ... m/s

c) Complete the graph in **Figure 2** to show Mandip's run from her classroom to the playing field (with her boots).

Studying distance-time graphs — it's a learning curve...

Being able to get information from graphs is a great skill to have in science. And both distance-time and velocity-time graphs are like wise old grandparents — full of information. Make sure you know how to read them properly.

Newton's First and Second Laws

When it comes to forces and acceleration, I'm not one to lay down the law. But Newton is...

Warm-Up

The definitions of Newton's First and Second Laws are as follows:

Newton's First Law: If the resultant force acting on an object is zero and:

1) the object is stationary — it remains stationary.
2) the object is moving at a given velocity — it continues to move at that same velocity.

Newton's Second Law: The acceleration of an object is proportional to the resultant force acting on the object and inversely proportional to the mass of the object.

What happens if the resultant force acting on an object is non-zero? Tick one.

☐ The object remains stationary. ☐ The object accelerates.

Write Newton's Second Law as an equation:

....

Q1 Are the forces acting on the object in the following situations balanced? Circle the correct answer.

a) A cricket ball slowing down as it rolls along a field.

Balanced Not balanced

b) A motorbike going round a roundabout at a steady 20 mph.

Balanced Not balanced

c) An ice skater moving in a straight line at a speed of 9.89 m/s for 3 seconds.

Balanced Not balanced

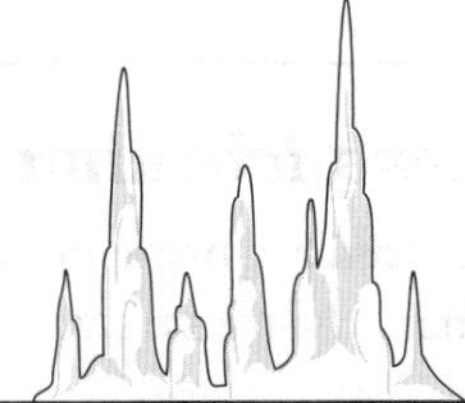

Q2 **Table 1** shows the mass and maximum driving force of four cars.

All four cars line up at the start of a race. When a pistol is fired, all the drivers apply their car's maximum driving force.

You can assume there are initially no forces opposing the motion of the cars and the mass of each driver is the same.

Tick the box next to the car in **Table 1** that has the highest acceleration when the pistol is fired.

Table 1

	Car	Mass (kg)	Maximum driving force (N)
☐ **A**	Newton 9000	800	4160
☐ **B**	Einstein 6i	1560	1100
☐ **C**	CGP TT	950	3090
☐ **D**	Heisenberg LM	2000	10 600

Q3 A 20.0 g bird is hit by a gust of wind.
The resultant force acting on the bird at this moment is –0.057 N.

a) Suggest **two** ways the bird's motion could change due to the gust of wind.

1. ....

2. ....

b) Calculate the acceleration of the bird as it is hit by the gust of wind.

acceleration = ... m/s^2

You know the physics — now take the law(s) into your own hands...

Questions about Newton's First and Second Laws always involve the *resultant* force. Many different forces can act on an object but it is the overall, resultant force that changes an object's velocity and is proportional to the acceleration.

 ☐  ☐  ☐

Newton's Third Law and Investigating Motion

First the worst, second the best, third an equal force contest...

Warm-Up

The definition of Newton's Third Law is as follows:

Newton's Third Law:	When two objects interact, the forces they exert on each other are equal and opposite.

Q1 A man leans against a wall. Circle the correct underlined phrase to complete the following sentence.

The size of the force exerted by the man on the wall is **larger than / smaller than / the same as** the size of the force exerted by the wall on the man.

Q2 Aria goes to the gym and holds a barbell into the air.

Aria's hands exert a normal contact force on the barbell.
Which of the following forces is the equal and opposite force to this?
Tick **one** box.

☐ **A** The barbell exerting a gravitational force on Aria's hands.

☐ **B** The barbell exerting a normal contact force on Aria's hands.

☐ **C** The floor exerting a normal contact force on Aria's hands.

Q3 A ping pong player uses her bat to hit a ball with a force of 4.2 N.

a) What is the force the ball exerts on the bat at the moment of contact? Tick **one** box.

☐ −8.4 N ☐ −4.2 N ☐ 4.2 N ☐ 8.4 N

b) After the ball has been hit, the ball lands on the table.
Complete the paragraph below using the words and phrases in the box.

gravitational	opposite directions	the same direction
different	normal contact	equal

When the ball is lying on the table, the Earth exerts a gravitational force on the ball, and the ball exerts a .... force on the Earth.

These two forces are .... in size and act in ....

Q4 The acceleration of a trolley down a ramp can be measured using the setup in **Figure 1**.

Figure 1

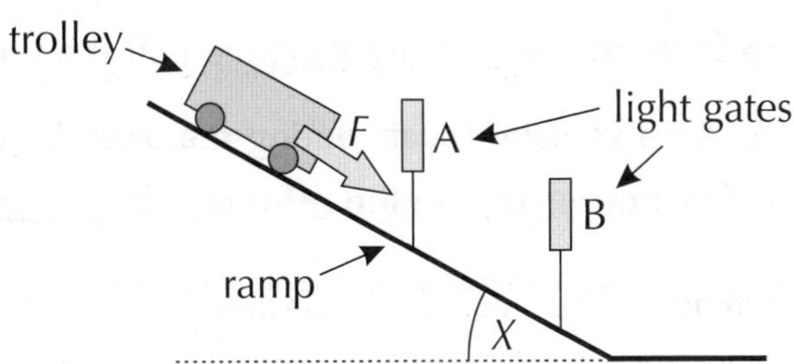

Assume the ramp is frictionless so that the only force acting parallel to the ramp is force *F*, pulling the trolley down. Increasing the angle of the ramp, *X*, increases force *F*.

Light gates are used to record the trolley's velocity at two different points on the ramp. Light gate A records the initial velocity, *u*, and light gate B records the final velocity, *v*, between the two points. From this, the acceleration, *a*, can be calculated.

a) Use the values in **Table 1** to complete the acceleration column in **Table 1**.

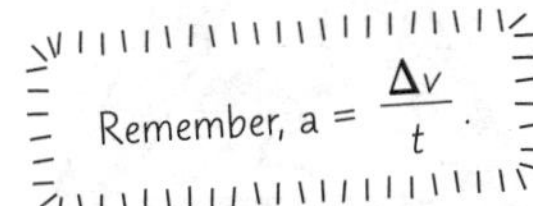

Table 1

X (°)	*F* (N)	*u* (m/s)	*v* (m/s)	*t* (s)	*a* (m/s²)
30	0.36	0.17	0.67	1.00	0.50
35	0.41	0.23	0.69	0.81	
40	0.46	0.26	0.75	0.77	0.64
45	0.51	0.34	0.81	0.66	

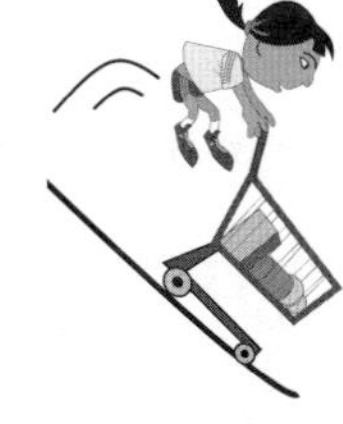

b) Use **Table 1** to plot a graph of force against acceleration on **Figure 2**. Draw a line of best fit.

c) Use your plot to determine the trolley's mass.

trolley mass = ... kg

Figure 2

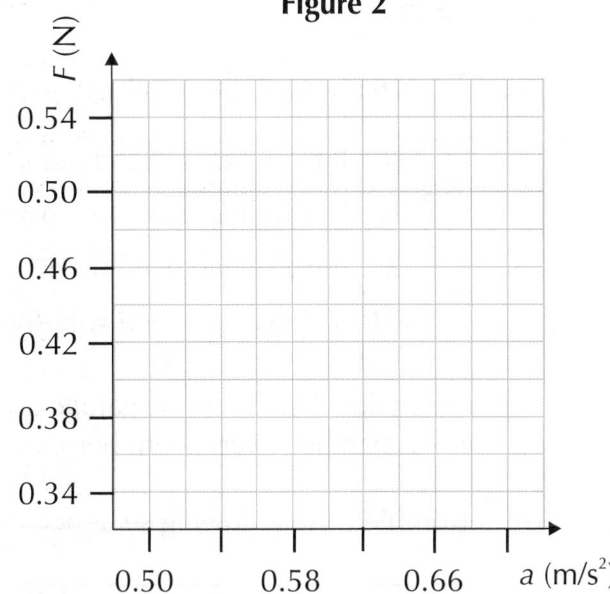

CGP Cinema presents: Newton's experiments — a proper motion picture...

Some variables (control variables) must be kept constant so that experiments produce reliable results. Here, when testing Newton's Second Law, one of mass, force or acceleration must be kept constant (as the control variable). F(=m)antastic.

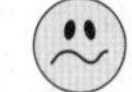 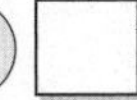  

Stopping, Thinking and Braking Distances

Stopping distance is the minimum distance a vehicle will travel when coming to a stop. Simple.

Warm-Up

The stopping distance is made up of the thinking distance and the braking distance.

The thinking distance is how far a vehicle travels during the driver's reaction time.
The braking distance is the distance taken to stop after the driver applies the brakes.

Tick any of the following distances that would be affected by wet roads.

☐ Thinking distance ☐ Braking distance ☐ Stopping distance

Q1 Hafsa is driving along the road and sees an ostrich 47 m ahead. She hits the brakes hard.

The car covers a distance of 29 m between her hitting the brakes and the car coming to a stop. Hafsa's thinking distance is 16 m.

a) What is Hafsa's braking distance? Circle the correct answer.

13 m 29 m 45 m

b) What is Hafsa's stopping distance? Circle the correct answer.

13 m 29 m 45 m

c) Tick **one** factor below that would have increased Hafsa's thinking distance.

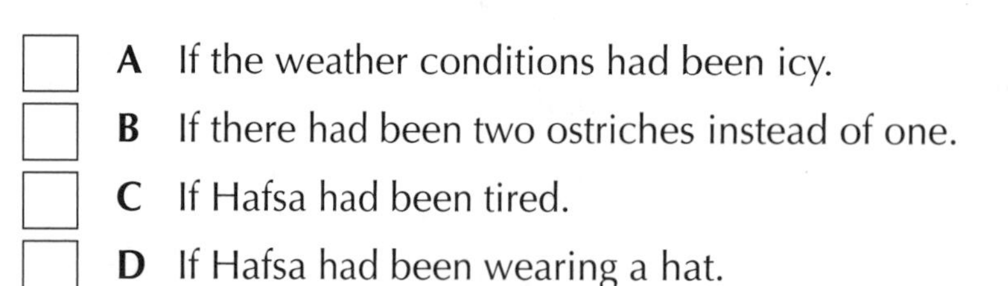

☐ **A** If the weather conditions had been icy.
☐ **B** If there had been two ostriches instead of one.
☐ **C** If Hafsa had been tired.
☐ **D** If Hafsa had been wearing a hat.

Hafsa had taken her car to the garage a week ago to get new brake pads. Her brakes had been very worn beforehand.

d) Complete the following sentence by circling the correct phrase.

If the ostrich had been on the road in the same conditions two weeks ago then Hafsa would have been **more likely** / **less likely** to hit it.

Q2 Stan is driving along a motorway when he sees a hazard on the road in front of him. He hits the brakes.

a) When the brakes of Stan's car are used, they heat up. Fill in the passage using some of the words below to explain why.

friction	tension	thermal	potential	kinetic

When the brake pedal is pushed, the brake pads are pressed onto the wheels causing .... This causes work to be done by the brakes, transferring energy from the .... energy stores of the wheels to the .... energy stores of the brakes. This causes the brakes to heat up.

b) The car has a large deceleration. Tick **two** reasons why this could be dangerous.

- [ ] **A** It could cause Stan to lose control of the car.
- [ ] **B** It could cause the car's engine to explode.
- [ ] **C** It could cause the brakes to cool down, increasing the stopping distance.
- [ ] **D** It could cause the brakes to overheat, meaning they won't work as well.

Q3 The speed limit in a village has recently been reduced from 40 mph to 30 mph to improve safety in the village.

One resident says, "A driver's reaction time is the same at both 30 mph and 40 mph which means the thinking distance of a car is the same at both speeds. So changing the speed limit doesn't make the village safer."

Explain why the resident is wrong.

....

....

....

....

....

....

It's not too difficult if you brake it all down...

Stopping distance = thinking distance + braking distance. Got it? Good. Make sure you know what factors affect the stopping distance, thinking distance and braking distance, and how they affect them. Then drive safely to the next page.

PRACTICAL

Reaction Times

Think fast! This page is all about reaction time experiments...

Warm-Up

Reaction times are different for different people.

State a typical value for a person's reaction time.

....

Q1 Scientists A and B each do a computer test to find their reaction times. The test requires them to click the mouse as quickly as they can when an image pops up on the screen. They each do the experiment three times. **Table 1** shows their results.

Table 1

	Reaction time (s)		
Scientist A	0.80	0.76	0.69
Scientist B	0.32	0.22	0.25

a) Use **Table 1** to find the average reaction time of each scientist.

A's reaction time = ... s

B's reaction time = ... s

Scientist C wants to test the effects of alcohol on reaction times. She completes the computer test three times and then has a glass of wine. She then takes the test three times again. **Table 2** shows her results.

Table 2

Attempt	Reaction time before alcohol (s)	Reaction time after alcohol (s)
1	0.32	0.58
2	0.30	0.54
3	0.29	0.62

b) Explain how the information from **Table 2** suggests that it is not safe to drive after drinking alcohol.

Remember, thinking distance is affected by reaction time...

....

....

....

....

Get up to speed — learn about reaction times...

Tiredness is another thing that can affect your reaction times. You could try using that as an excuse next time you fancy a lie in — "I'd love to get up and go to school, but I think I need more sleep. You know, just in case I run into any bears on the way to school and need to react quickly to the situation." Hmm... yeah ok, that excuse might need some work...

Waves — The Basics

Make sure you get to grips with wave basics before jumping in at the deep end...

Warm-Up

<u>All waves</u> have an amplitude, wavelength, frequency and period.

Use <u>two</u> of the words below to fill in the sentences about waves.

Frequency	Amplitude	Period	Wavelength

The .... is the number of waves passing a point each second.

The time it takes for a wave to complete a full cycle is called the ....

The <u>wave equation</u> applies to all waves.
It links the wave speed to wavelength and frequency.

<u>State</u> the wave equation below:

....

Q1 A student measures a sound wave. They draw a displacement-distance graph for the sound wave. This is shown in **Figure 1**.

Figure 1

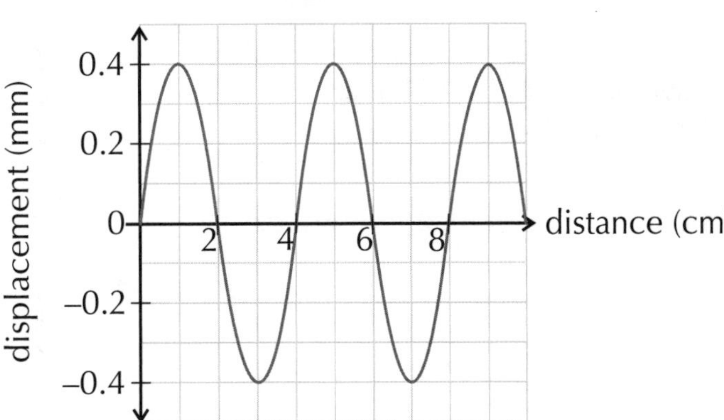

a) Did the student measure a transverse or longitudinal wave?

....

b) Tick **two** options that correctly describe the wave drawn in **Figure 1**.

- [ ] **A** The amplitude in **Figure 1** is 0.4 mm.
- [ ] **B** The amplitude in **Figure 1** is 0.8 mm.
- [ ] **C** The wavelength in **Figure 1** is 4 cm.
- [ ] **D** The wavelength in **Figure 1** is 5 cm.

Q2 A bird is sat on the surface of a pond. A water wave in the pond makes the bird bob up and down three times every second.

a) What is the frequency of the bird's bobbing?
Tick the correct answer.

☐ 0.3 Hz ☐ 3 Hz ☐ 5 Hz ☐ 9 Hz

b) The bird's bobbing frequency is the same frequency as the water wave.
Calculate the period of the water wave.

period = ... s

c) There is a small piece of bread one wavelength away from the bird.
The water wave travels at 0.5 m/s. Calculate the distance between the bird and the small piece of bread.

distance = ... m

d) The water wave travels to the edge of the pond. Does the bird travel to the edge of the pond with the water wave? Explain your answer.

....

....

Q3 A radio wave is travelling at 3.0×10^8 m/s.
It has a period of 1.5×10^{-10} s.

a) Calculate the frequency of the radio wave.

frequency = ... Hz

b) Calculate the wavelength of the radio wave.

wavelength = ... m

Put down your bucket and spade. The waves are here...

The wave equation applies to all waves. It doesn't matter if the wave is transverse or longitudinal. It doesn't matter if it's a light or sound wave. All that matters is that the equation applies to them all — so it is definitely worth learning.

Investigating Waves and Refraction

Time to wave hello to some experiments, and the weird world of refraction...

Warm-Up

The set-up below can be used to measure the wave speed in a string.

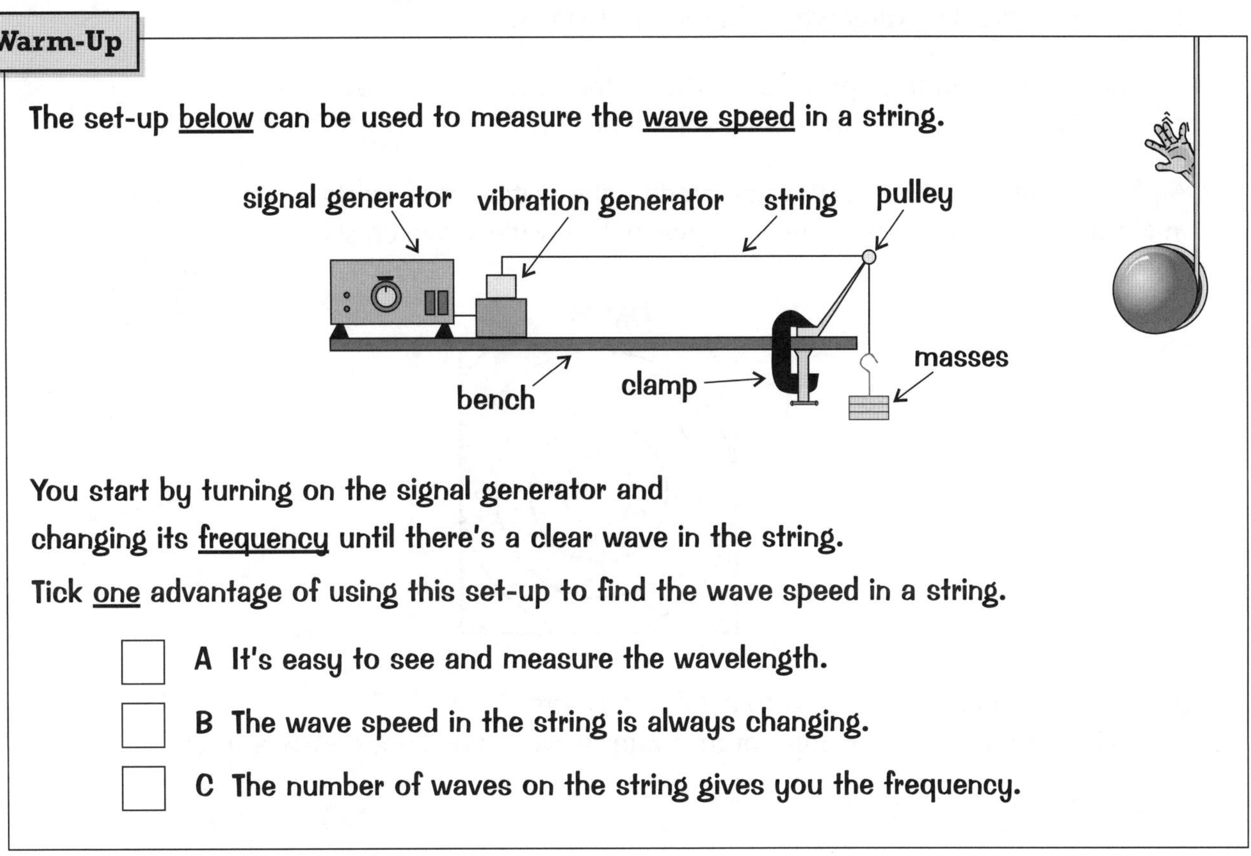

You start by turning on the signal generator and changing its frequency until there's a clear wave in the string.

Tick one advantage of using this set-up to find the wave speed in a string.

- [ ] **A** It's easy to see and measure the wavelength.
- [ ] **B** The wave speed in the string is always changing.
- [ ] **C** The number of waves on the string gives you the frequency.

Q1 A ray of red light is shone at a glass bottle from the air. The light ray refracts when it crosses the boundary from the air to the glass bottle. The ray diagram for this is shown in **Figure 1**. The ray diagram is incomplete.

Figure 1

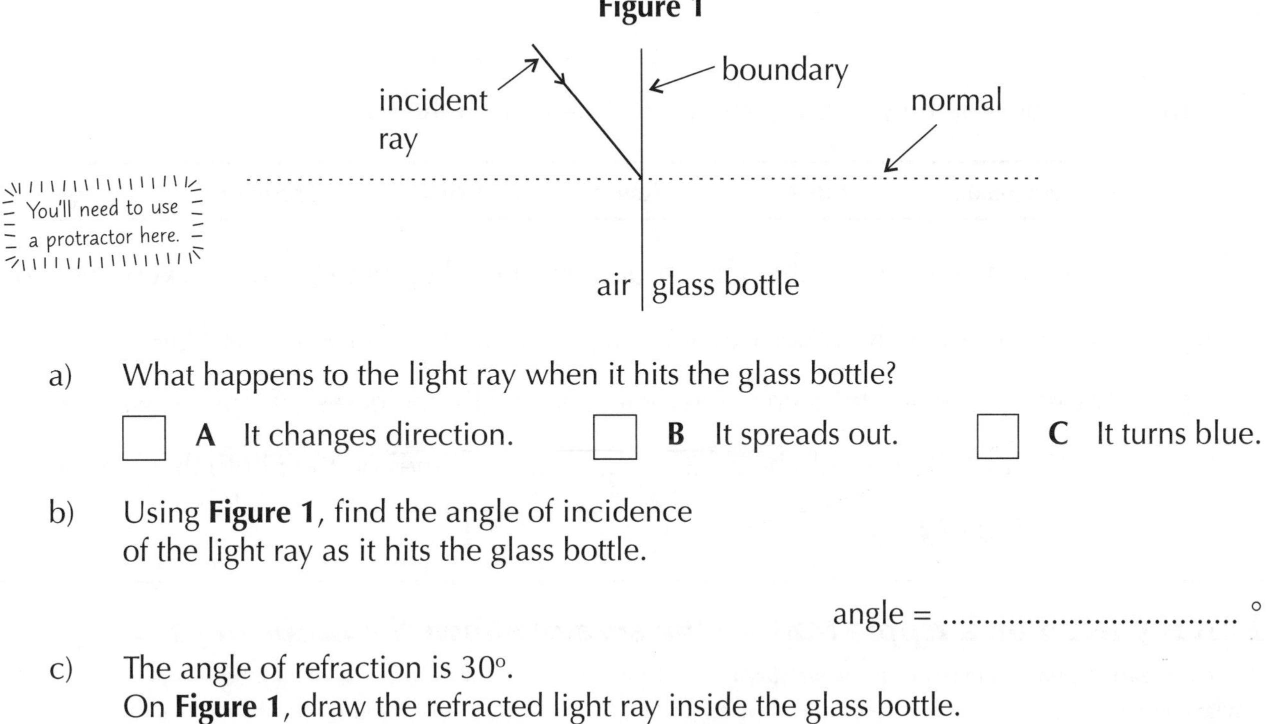

a) What happens to the light ray when it hits the glass bottle?

- [ ] **A** It changes direction.
- [ ] **B** It spreads out.
- [ ] **C** It turns blue.

b) Using **Figure 1**, find the angle of incidence of the light ray as it hits the glass bottle.

angle = ... °

c) The angle of refraction is 30°.
On **Figure 1**, draw the refracted light ray inside the glass bottle.

PRACTICAL

Q2 A student experiments with a ripple tank. They drop one drop of water every half second onto the middle of the ripple tank. This creates circular water waves in the ripple tank.

A lamp is shone on the ripple tank. The water wave crests can be seen as shadows on a screen below.

A wave crest is the highest point of a wave.

A photo is taken of the shadows created by the water wave crests — this is shown in **Figure 2**. The lines show the shadows of the water wave crests.

Figure 2

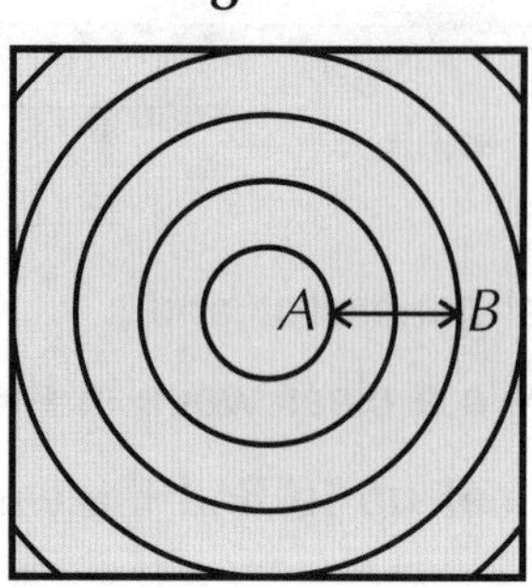

a) i) The distance from crest *A* to crest *B* is 16 cm.
Determine the wavelength of the water waves. Give your answer in cm.

wavelength = ... cm

ii) Every second, two water waves reach the edge of the ripple tank.
Calculate the wave speed of the water waves.

wave speed = ... m/s

b) Use some of the words below to fill in the gaps in the paragraph.

decreases	smaller	fewer	more	increases	bigger

If the student increases the number of drops they drop onto the water every second, the frequency of the water waves .... The speed of the water waves will remain the same, and their wavelength decreases. If a photo was taken in this case, there would be .... lines on the photo than on the one in **Figure 2**.

I hit my head on a ripple tank — the screen showed a brain wave...

It's worth learning how to measure the wave speed in a string and of a water wave. Thankfully, both experiments follow similar methods. You find the frequency and wavelength first, then calculate the speed using the wave equation.

Electromagnetic Waves

Buy one get the other six free — a one-off bargain deal for the entire electromagnetic spectrum...

Warm-Up

There are seven different types of electromagnetic wave.
They form a continuous spectrum.

The different types of electromagnetic wave are shown below.
Put them in order, from the lowest to the highest frequency, in the table below.

Radio waves   Infrared   X-rays   Gamma rays   Ultraviolet   Microwaves

			VISIBLE LIGHT			

Low frequency ⟶ High frequency

Q1 A scientist is investigating the electromagnetic radiation emitted by cobalt atoms. Use some of the words below to complete the paragraph about cobalt atoms.

electrons	narrow	visible light	gamma rays	radio waves	wide

Cobalt can produce electromagnetic waves with a .... range of frequencies.

In a cobalt atom, ... can move between energy levels by absorbing or emitting electromagnetic waves. Changes in the cobalt atom's nucleus can cause the atom to emit ....

Q2 Electromagnetic waves transfer energy from their source to an absorber.

For each of the following examples of energy transfer, name the source of the wave, the absorber of the wave and the type of electromagnetic wave involved.

a) A toaster heating a slice of bread.

Source [ ] ⟶ Electromagnetic wave [ ] ⟶ Absorber [ ]

b) A person watching television.

Source [ ] ⟶ Electromagnetic wave [ ] ⟶ Absorber [ ]

Q3 An alien uses electromagnetic waves to charge its spaceship.

a) When charging, the spaceship absorbs the type of electromagnetic wave with the shortest wavelength. Which type of electromagnetic wave does the alien's spaceship absorb when charging? Tick **one** box.

Gamma rays ☐ Microwaves ☐

Ultraviolet ☐ Visible light ☐

b) The spaceship detects an electromagnetic wave from space. The wave has a frequency lower than all X-ray frequencies and a wavelength shorter than all infrared wavelengths.

Circle the **two** types of electromagnetic wave that this wave could be.

Ultraviolet **Infrared** **Microwaves** **X-rays**

Gamma rays **Visible light** **Radio waves**

Q4 A teacher has a laser that emits visible light. Draw a circle to show whether each of the following statements about the light from the laser is true or false.

The light from the laser can be detected by most human eyes.	True	False
The light from the laser is a longitudinal wave.	True	False
The light from the laser has a shorter wavelength than all microwaves.	True	False
The light from the laser transfers energy to an absorber.	True	False

Q5 The Sun gives out all types of electromagnetic waves. Visible light waves take 8 minutes and 20 seconds to reach the Earth from the Sun. How long does it take infrared waves to reach the Earth from the Sun? Explain your answer.

....

....

....

....

The electromagnetic spectrum — the seven wonders of the physics-world...

A place with heat, light or even a hip-hop radio station is a place where electromagnetic waves are transferring energy. Transfer your energy into learning all the electromagnetic waves, from low to high frequency. It's a handy thing to know.

EM Waves and Their Uses

As you'll see, electromagnetic waves can do loads...

Warm-Up

All <u>seven</u> types of electromagnetic (EM) waves are <u>useful</u> for different sorts of <u>jobs</u>.

<u>Fibre optic</u> cables carry data over long distances to telephones and computers.
What type of EM wave below can be used to transmit the data? <u>Tick</u> the correct answer.

- [ ] Visible light
- [ ] Microwaves
- [ ] X-rays
- [ ] Ultraviolet

X-rays and gamma rays can be used in <u>hospitals</u>.
State <u>two</u> ways gamma rays can be used in hospitals.

1. ....

2. ....

Q1 Dom thinks his arm is broken, so he visits a doctor.

a) The doctor is thinking of using EM waves to see if Dom's arm is broken.
Draw a circle around the type of EM wave the doctor should use.

Infrared   Gamma rays   Radio waves   X-rays

b) Dom finds out that his arm is not broken.
In celebration, he goes to the beach and gets a suntan.
What type of EM waves has Dom's skin absorbed to get a suntan? Tick the correct answer.

- [ ] **A** Gamma rays
- [ ] **B** Visible light
- [ ] **C** Ultraviolet
- [ ] **D** Radio waves

c) While at the beach, Dom buys an ice cream. The ice cream absorbs infrared radiation from the Sun. Explain why this causes the ice cream to melt.

....

....

....

Q2 Microwave ovens use microwaves to cook food. An egg is cracked into a glass jug and placed inside a microwave oven for 45 seconds.

a) Explain why the egg will heat up but the glass jug will not.

b) Which EM wave is also commonly used to cook eggs?
Circle the correct answer.

Microwaves   Gamma rays   Infrared   Ultraviolet

X-rays   Visible light   Radio waves

Q3 Infrared cameras detect infrared radiation and create a picture.

a) i) An infrared camera is used to take a picture of an elephant close to a tree.
State whether the tree or the elephant will show up more brightly on the image.

ii) Explain your answer.

b) Firefighters sometimes use infrared cameras to look at a building that is on fire. Suggest an advantage of using an infrared camera in this case.

Radio waves — it's what all the sailors listen to...

Bet you never thought EM waves could be so useful. It's thanks to them we can cook eggs in 45 seconds. Although they usually taste nasty. Nothing is worse than a tasteless egg — not even GCSE Science. Anyway, let's crack on...

Investigating IR Radiation

PRACTICAL

Find a Leslie cube. Dig out that infrared detector. Things are heating up...

Warm-Up

The amount of infrared (IR) radiation an object emits (gives out) depends on its temperature and on its surface. This can be investigated with a Leslie cube.

Use some of the following words to fill in the gaps in the paragraph below:

colour	temperature	equal	different	hardness

Four side faces of a Leslie cube vary in roughness, shininess and .... When a Leslie cube is filled with boiling water, all four faces have the same temperature and emit .... amounts of IR radiation.

Q1 A Leslie cube has four different surfaces. One surface has shiny black paint on it. Another has dull black paint. The third has a layer of dull copper. The fourth has a layer of shiny silver.

The cube is filled with boiling water. After one minute, an IR radiation detector measures the intensity of IR radiation emitted from each surface of the cube.

During the measurements, the IR radiation detector is held at the same distance from each surface of the cube.

Table 1 shows the intensity of IR radiation measured for each surface.

Table 1

Surface	Shiny black paint	Dull black paint	Dull copper	Shiny silver
IR intensity (W/m^2)	13.2	18.9	7.4	3.6

a) Look at **Table 1**. Which surface is emitting the most IR radiation?

....

b) Suggest why the IR radiation detector is held at the same distance from each surface of the cube.

....

....

....

That old infrared detector — it's been around the block a few times...

Folks, here's a handy hint from my friend Rebekah... don't touch a Leslie cube in action — its sides are all sizzling hot. Remember, although the sides are all at the same temperature, the intensity of IR radiation emitted from each side varies.

Dangers of Electromagnetic Waves

- DANGER — DANGER — DANGER — DANGER — DANGER — DANGER — DANGER — DANGER -

Warm-Up

Some EM waves can be harmful and cause damage to your body.

Circle whether the following statements are true or false:

Ultraviolet waves can cause skin to age slower than it should.	True	False
Ultraviolet waves, X-rays and gamma rays can all increase the risk of cancer.	True	False
Radiation dose is a measure of the risk of harm to the body from an exposure to radiation.	True	False
The radiation from one X-ray scan carries a risk of causing cancer.	True	False

Q1 Adya has a bad chest infection. Doctors are deciding whether to give her a standard X-ray scan or a CT scan. The radiation dose for each is shown in **Table 1**.

Table 1

Scan	Radiation dose (mSv)
Chest X-ray	0.1
CT-Chest	7

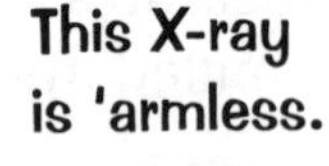

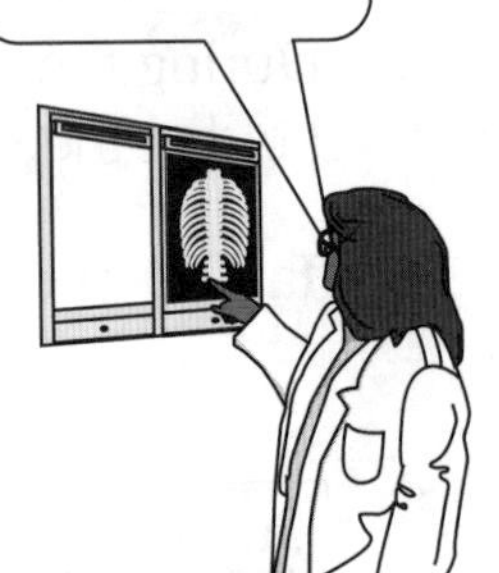

a) Using **Table 1**, calculate the increase in the risk of harm if Adya has a CT scan instead of an X-ray scan.

Adya's increase in risk of harm = ... times higher

b) Adya may also need an X-ray scan on her hand. Assuming the X-rays used for both scans are the same, would you expect a hand X-ray scan to have a higher or lower radiation dose than a chest X-ray scan? Explain your answer.

....

....

....

Having an X-ray scan in winter — it can chill you to the bone...

Most people will have an X-ray scan at some point. One every year or so is pretty low risk, so they're not worth worrying about too much. However, X-rays are harmful EM waves, so a scan is never going to be totally risk-free.

Permanent and Induced Magnets

A lot of people think magnetism is magic. I'll let you in on a secret though — it's just physics.

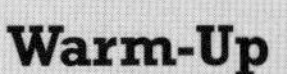

All magnets create magnetic fields.

The magnetic field strength changes depending on where you are around a magnet.
The stronger the magnetic field strength between two magnets, the stronger the attraction or repulsion between the magnets.

Three pairs of magnets are shown below.
Tick the box next to any of the pairs that will attract each other.

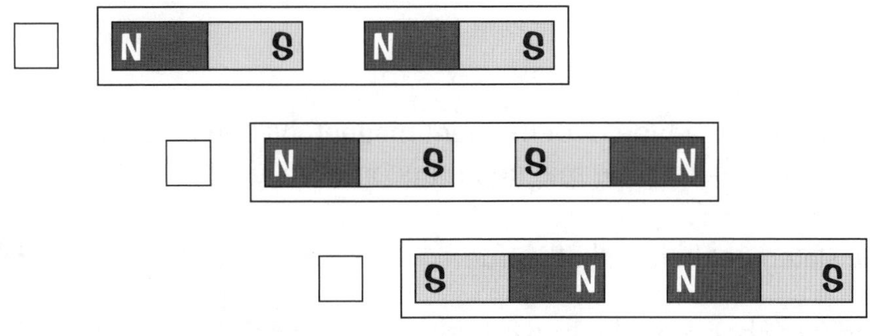

Q1 The bar magnet shown in **Figure 1** has three paper clips around it.
All the paper clips are attracted to the bar magnet.
Circle the paper clip that feels the strongest attraction.

Figure 1

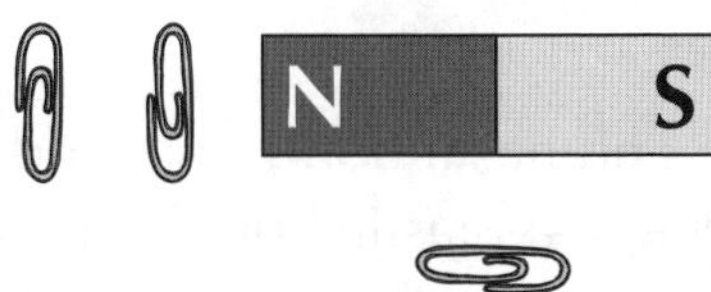

Q2 Four compasses are placed around a bar magnet, as shown in **Figure 2**.
The compasses are shown as circles. One of the compasses has an arrow drawn in it to show the direction that the compass points in.

Draw an arrow in the other three compasses to show the direction they point in.

Figure 2

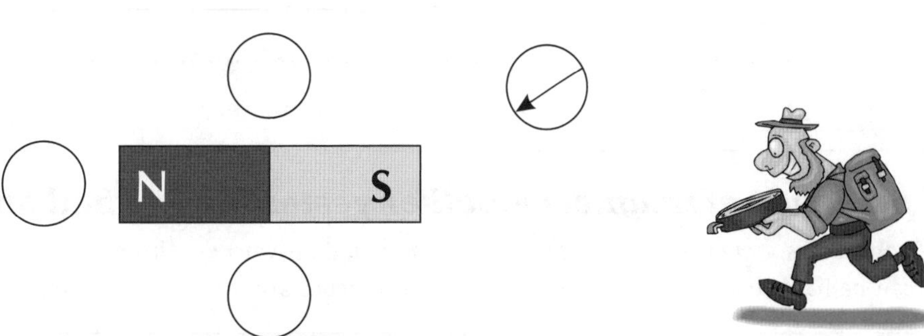

Q3 A student has a bar magnet and a block of metal. When she brings the north pole of the bar magnet close to a point on the block of metal, they repel each other.

a) What is the block of metal? Tick **one** box.

☐ a bar magnet ☐ an induced magnet

b) Use **one** of the phrases in the box to complete the sentence below.

will attract	will repel	won't attract or repel

The bar magnet is rotated so its south pole is brought close to the same point on the block as before. The bar magnet and the block .... each other.

Q4 A horseshoe magnet is a type of permanent magnet. When an iron nail is brought close to a horseshoe magnet, it becomes magnetic.

Tick the **one** statement below that is true.

☐ **A** The force between the horseshoe magnet and the iron nail will be repulsive.

☐ **B** The iron nail will lose most of its magnetism when the horseshoe magnet is removed.

☐ **C** The horseshoe magnet will lose most of its magnetism when the iron nail is removed.

☐ **D** The horseshoe magnet will become an induced magnet.

Q5 A needle is magnetised so that one end has a north pole and the other end has a south pole. The needle is stuck to a plastic bottle cap and floated on the surface of some water in a tank. The needle is not close to any magnets.

After floating for a short time, the needle points north.
Explain what this suggests about the Earth's core.

....

....

....

....

I put loads of magnets together yesterday — I had an absolute field day...

Magnetism is a great part of physics and can be found anywhere — from fridge magnets to the centre of the Earth. It can be hard to get your head around, but with practice and hard work it should become a rather *attractive* topic.

Electromagnetism

Electromagnetism is the name, creating magnetic fields from electric currents is the game.

Warm-Up

When a current flows through a wire, a magnetic field is created <u>around</u> the wire.
Use <u>some</u> of the words in the box to complete the sentences <u>below</u>:

increases	stays the same	stronger	decreases	weaker

The magnetic field strength around a wire ....
when the current through the wire is increased. The magnetic
field is .... closer to the wire.

Q1 A student bends a wire to form a solenoid. The solenoid is supplied with a current so that it creates a magnetic field. The student then puts an iron key in the middle of the solenoid. This is shown in **Figure 1**.

Figure 1

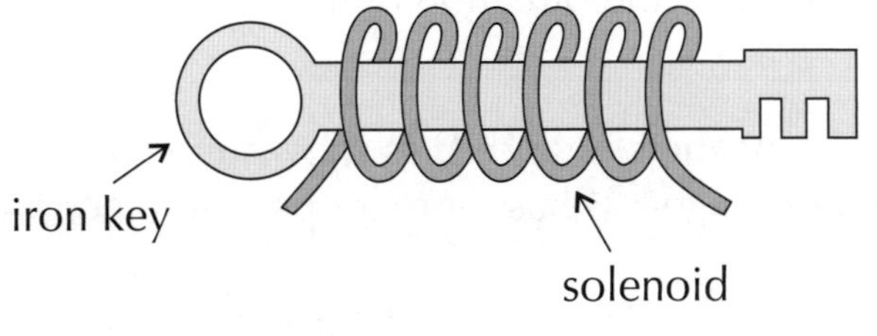

a) Circle the correct word in the sentence below to describe how the iron key affects the solenoid's magnetic field.

The strength of its magnetic field is **increased / decreased / unaffected** .

b) Circle the letter under **one** diagram below that correctly shows the magnetic field lines inside the middle of the iron key.

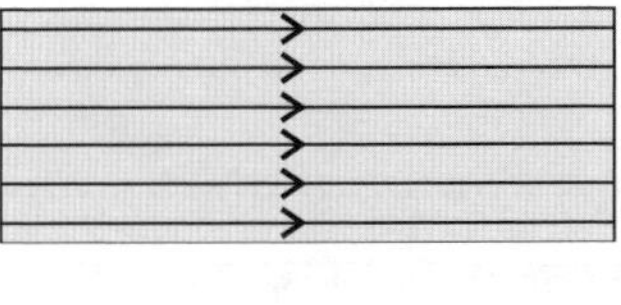

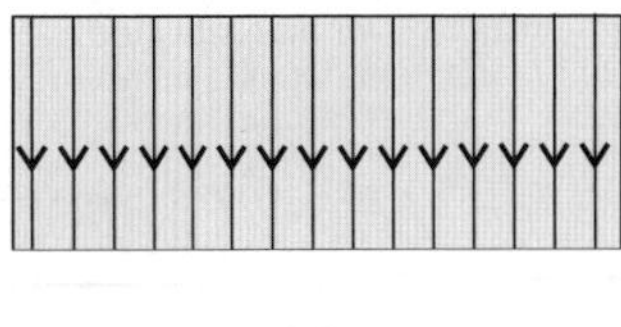

A **B** **C**

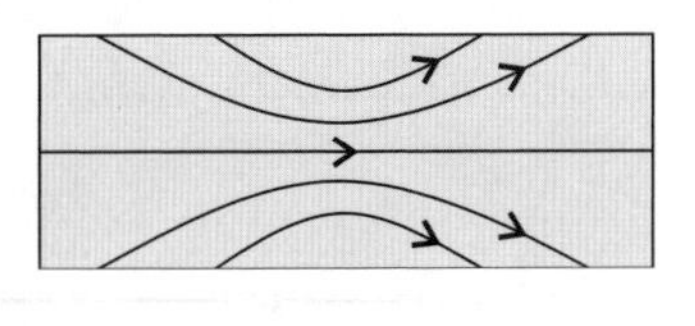

c) The current supplied to the solenoid is decreased.
State how the magnetic field strength inside the iron key changes.

....

PRACTICAL

Q2 Rani passes a loop of wire through a piece of card, as shown in **Figure 2**. The wire is then connected to a circuit. The circuit is controlled with a switch. When the switch is closed, a current flows through the wire.

Figure 2

current when switch closed

piece of card

loop of wire

a) Which of the following describes what happens when the switch is closed? Tick **one**.

☐ **A** A magnetic field is created by the wire.

☐ **B** A magnetic field is created by the piece of card.

☐ **C** A magnetic field is created by both the wire and the card.

b) Rani sprinkles some iron filings onto the card. When the switch is closed, the iron filings follow the pattern of the magnetic field created in a).
On **Figure 2**, sketch the pattern the iron filings make.
Include arrows to show the direction of the magnetic field.

Q3 Josh makes two electromagnets, A and B, as shown in **Figure 3**. They are both 5 cm long. Electromagnet B has more loops than electromagnet A.

Figure 3

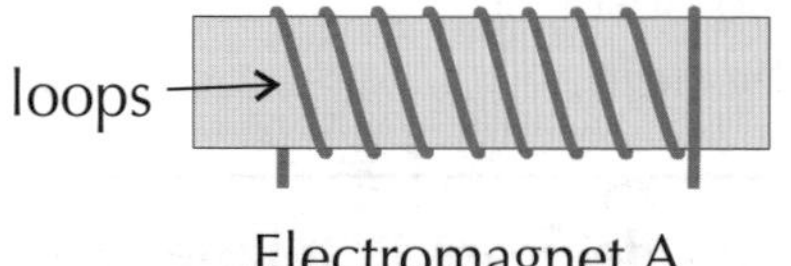

Electromagnet A

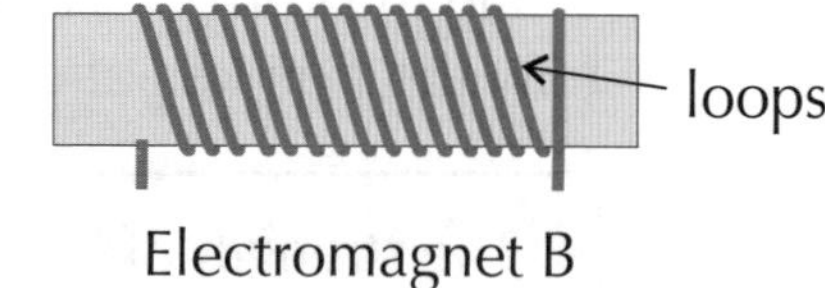

Electromagnet B

Both electromagnets are supplied with the same current.
Explain why electromagnet B creates a stronger magnetic field than electromagnet A.

Hint: each loop on the electromagnet produces its own magnetic field.

....

....

....

....

....

I have an old electromagnet that doesn't work — it's rotten to the core...

A stronger magnetic field is created when current flows through a solenoid rather than a straight wire. This is because solenoids, like all great roller coasters, are loopy. Each loop helps to increase the overall magnetic field strength.

The Periodic Table

Relative atomic mass (top number) — Atomic (proton) number (bottom number)

Periods	Group 1	Group 2											Group 3	Group 4	Group 5	Group 6	Group 7	Group 0
1								1 H Hydrogen 1										4 He Helium 2
2	7 Li Lithium 3	9 Be Beryllium 4											11 B Boron 5	12 C Carbon 6	14 N Nitrogen 7	16 O Oxygen 8	19 F Fluorine 9	20 Ne Neon 10
3	23 Na Sodium 11	24 Mg Magnesium 12											27 Al Aluminium 13	28 Si Silicon 14	31 P Phosphorus 15	32 S Sulfur 16	35.5 Cl Chlorine 17	40 Ar Argon 18
4	39 K Potassium 19	40 Ca Calcium 20	45 Sc Scandium 21	48 Ti Titanium 22	51 V Vanadium 23	52 Cr Chromium 24	55 Mn Manganese 25	56 Fe Iron 26	59 Co Cobalt 27	59 Ni Nickel 28	63.5 Cu Copper 29	65 Zn Zinc 30	70 Ga Gallium 31	73 Ge Germanium 32	75 As Arsenic 33	79 Se Selenium 34	80 Br Bromine 35	84 Kr Krypton 36
5	85 Rb Rubidium 37	88 Sr Strontium 38	89 Y Yttrium 39	91 Zr Zirconium 40	93 Nb Niobium 41	96 Mo Molybdenum 42	98 Tc Technetium 43	101 Ru Ruthenium 44	103 Rh Rhodium 45	106 Pd Palladium 46	108 Ag Silver 47	112 Cd Cadmium 48	115 In Indium 49	119 Sn Tin 50	122 Sb Antimony 51	128 Te Tellurium 52	127 I Iodine 53	131 Xe Xenon 54
6	133 Cs Caesium 55	137 Ba Barium 56	139 La Lanthanum 57	178 Hf Hafnium 72	181 Ta Tantalum 73	184 W Tungsten 74	186 Re Rhenium 75	190 Os Osmium 76	192 Ir Iridium 77	195 Pt Platinum 78	197 Au Gold 79	201 Hg Mercury 80	204 Tl Thallium 81	207 Pb Lead 82	209 Bi Bismuth 83	[209] Po Polonium 84	[210] At Astatine 85	[222] Rn Radon 86
7	[223] Fr Francium 87	[226] Ra Radium 88	[227] Ac Actinium 89	[261] Rf Rutherfordium 104	[262] Db Dubnium 105	[266] Sg Seaborgium 106	[264] Bh Bohrium 107	[277] Hs Hassium 108	[268] Mt Meitnerium 109	[271] Ds Darmstadtium 110	[272] Rg Roentgenium 111	[285] Cn Copernicium 112	[286] Uut Ununtrium 113	[289] Fl Flerovium 114	[289] Uup Ununpentium 115	[293] Lv Livermorium 116	[294] Uus Ununseptium 117	[294] Uuo Ununoctium 118

The Lanthanides (atomic numbers 58-71) and the Actinides (atomic numbers 90-103) are not shown in this table.

Physics Equations List

Physics Equations List

Here are some equations you might find useful when you're working through the Physics sections of this book — you'll be given these equations in the Physics exams.

Topic P1 — Energy

$E_e = \frac{1}{2}ke^2$	elastic potential energy = 0.5 × spring constant × (extension)^2
$\Delta E = mc\Delta\theta$	change in thermal energy = mass × specific heat capacity × temperature change

Topic P3 — Particle Model of Matter

$E = mL$	thermal energy for a change of state = mass × specific latent heat

Topic P5 — Forces

$v^2 - u^2 = 2as$	(final velocity)^2 – (initial velocity)^2 = 2 × acceleration × distance

Topic P6 — Waves

$$\text{period} = \frac{1}{\text{frequency}}$$

Formula Triangles

It's pretty important to learn how to put a formula into a triangle. There are two easy rules:

1) If the formula is "$A = B \times C$" then A goes on the top and $B \times C$ goes on the bottom.
2) If the formula is "$A = B \div C$" then B must go on the top (because that's the only way it'll give "B divided by something") — and so A and C must go on the bottom.

Two Examples:

$F = ma$ turns into:

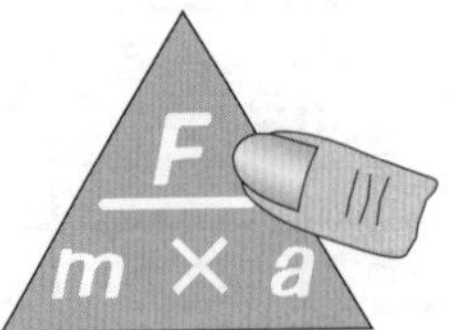

$P = E \div t$ turns into:

How to use them: Cover up the thing you want to find and write down what's left showing.

EXAMPLE: To find m from the first one, cover up m and you get $\frac{F}{a}$ left showing, so "$m = \frac{F}{a}$".

Using Formulas — the Three Rules:

1) Find a formula which contains the thing you want to find as well as the other things that you've got values for. Change that formula into a formula triangle.
2) Check your values are in the right units, then stick the numbers in.
3) Work out the answer and check that it is sensible.